Marek J. Murawski
Marek Ryś

Junkers Ju 88

vol. II

KAGERO

Junkers Ju 88 vol. II • Marek J. Murawski, Marek Ryś • First edition • LUBLIN 2014

 • **ISBN 978-83-64596-27-8**

Editing: **Marek J. Murawski** • Translation: **Piotr Kolasa** • Cover artwork: **Marek Ryś** • Visualisation 3D: **Marek Ryś** • Photos: **Marek J. Murawski's archive** • Design: **KAGERO STUDIO, Łukasz Maj**

Oficyna Wydawnicza KAGERO
Akacjowa 100, Turka, os. Borek, 20-258 Lublin 62, Poland, phone/fax: (+48) 81 501 21 05
www.kagero.pl • e-mail: kagero@kagero.pl, marketing@kagero.pl
w w w . k a g e r o . p l
Distribution: **Oficyna Wydawnicza KAGERO**

The Phony War - September 3, 1939 - April 8, 1940

Soon after the full-scale production of the Ju 88 A-0 version had been launched ten examples of the bomber were delivered to Erprobungsstelle Rechlin on March 27, 1939. On the following day Erprobungskommando 88[1] became operational under the command of Hptm Helmut Pohle. The twelve crews forming the new unit were detached from 1./KG 55.

The main task of Erprobungskommando 88 was to perform operational trials of the new bomber. The first sorties revealed numerous issues with speed brakes actuation system, which on several occasions got stuck in the open position seriously affecting the aircraft's stability and limiting its speed.

After operational testing program had been completed, Erprobungskommando 88 became I./KG 25 in August 1939 with Hptm Helmut Pohle as the unit's CO. The group consisted of only two Staffel: 1./KG 25 and 2./KG 25.

When Britain and France formally declared war on Germany on September 3, 1939 1./KG 25 operated from Jever air base in northern Germany as part of 3. Fliegerdivision (3rd Air Division),

One of the first examples of the Junkers Ju 88 A-0 received by Erprobungskommando 88. Note the four-bladed propellers, which were replaced in later models, beginning with the A-1, with a three-bladed unit.

I./KG 25 was established in August 1939 at Jever, under the command of Hptm. Helmut Pohle. On September 22, 1939 the group was re-named I./KG 30.

Junkers Ju 88 A-0, D4+AH flown by the CO of 1./KG 30, the fall of 1939. Note the unit's insignia under the cockpit.

which was in turn subordinated to Luftflotte 2 (2nd Air Fleet). The unit had twelve aircraft on strength, all fully combat ready. Members of 2./KG 25 were at that time still undergoing operational flying training at Greifswald air base.

Two days after the outbreak of war in the West, on September 5, 1939, 10. Fliegerdivision was formed in Hamburg under the command of Gen.Lt. Hans Ferdinand Geisler. Maj. Martin Harlinghausen was appointed the Division's chief of staff. The unit was tasked with attacks on the Royal Navy vessels in the North Sea and bombing British naval installations. However, the assets available to Gen.Lt. Geisler (Heinkel He 111 bombers from KG 26 and just a dozen Ju 88s from 1./KG 25) were entirely inadequate to fulfill the unit's mission.

It is no wonder that after only a few days Gen.d.Fl. Helmut Felmy, commanding Luftflotte 2, decided that an experimental unit equipped with a new, unproven type could not be expected to perform such a difficult task and promptly ordered withdrawal of 1./KG 25 back to its home base at Greifswald, where 2./KG 25 was also based.

On September 22, 1939 I./KG 25 was renamed I./KG 30. At the same time the unit's strength was augmented by the addition of 3./KG 30. Hptm. Helmut Pohle remained at the helm of the Gruppe. In order to allow the Ju 88 crews to gain combat experience and to further test the design under operational conditions, Gen.d.Fl. Felmy authorized deployment of four bombers to an air base on the island of Sylt. The detachment was commanded by Lt. Walter Storp.

The Junkers Ju 88 crews detached from I./KG 30 had their baptism of fire on September 26, 1939. Two days before Kriegsmarine patrol vessels heavily damaged a British submarine HMS *Spearfish* near Horns Reef, just off the coast of Denmark. A day later the Royal Navy sent a rescue force consisting of battleships HMS *Rodney* and *Nelson*, battlecruisers *Hood* and *Renown* and an escort team of three cruisers and aircraft carrier HMS *Ark Royal*. The Home Fleet strike force was accompanied by a team of four cruisers and six destroyers, whose task was to provide direct cover for the damaged submarine.

At 10.45 on September 26, 1939 the crew of a Dornier Do 18 flying boat from Norderney-

The first combat unit to operate Junkers Ju 88 A-1 bombers was KG 30. Pictured here is one of the machines of Stab II./KG 30 (4D+DC).

based 2.(F)/106 spotted the British force during a routine patrol sortie. Having positively identified the force's vessels, the crew alerted the HQ of 10. Fliegerdivision. Gen.d.Fl. Geisler wasted no time and immediately ordered his crews to attack the British warships. At 12.50 nine Heinkel He 111 P-2 bombers from 1./KG 26 (commanded by Hptm. Vetter) launched from their base, followed by four Ju 88s from I./KG 30, which took off from Sylt at 13.00. Each of the Ju 88s, led by Lt. Walter Storp, carried two SC 500 bombs. As the Luftwaffe did not have any additional bombers available in the area, the opportunity to inflict serious damage on the British fleet was lost.

In an interview with a German war correspondent Gefr. Carl Francke[2] said: "On that day, at about 3.00 pm, I scored a direct hit dropping a 500-kg bomb on a British carrier. We were on combat alert at one of air bases when we received the news of a large British naval force leaving their bases. The message was radioed by the crew of a long-range maritime reconnaissance aircraft. Our orders were as follows: 'A large English surface force spotted in the North Sea. Priority target: aircraft carrier'.

My crew took off shortly before 13.00. As an experienced naval flyer I stayed at 500 m to avoid the cloud deck above us. Yet from that altitude I could still see far enough to be able to spot the enemy force.

As soon as I saw the English ships I climbed to 3,000 m to assume the most favorable attack position.

A group of young Junkers factory workers inspect a brand new Junkers Ju 88 A-1 parked at the factory airfield at Dessau.

Ju 88 A-1s awaiting delivery to KG 30.

This Junkers Ju 88 A-1 from 1./KG 30 is undergoing compass calibration. Note the unit insignia on the engine cowl – a white shield with Chamberlain's umbrella caught in the crosshairs.

As I punched through the overcast I realized the chances for a successful attack were practically nil, so I immediately aborted the run and climbed back into the clouds. The British escort ships opened heavy fire, but luckily my aircraft escaped any serious damage and I managed to climb back to altitude and set up for another try.

I commenced the second run from 2,700 m. Once again the target was completely obscured by the overcast. However, once I penetrated the cloud layer I saw an aircraft carrier directly in front of my nose.

The first bomb missed the British ship by about 20 meters, but the second found its mark and hit the carrier's starboard side. The British immediately responded with a ferocious barrage of anti-aircraft fire. I did not see the moment of the bomb's impact, but other crew members told me they had observed a thick, black cloud of smoke and a flash of fire coming from the carrier's starboard side.

Later on my colleagues from a maritime reconnaissance unit confirmed that the carrier, listing heavily and unable to maintain steady course, turned around and made for home port

followed by other vessels. The British force was spotted again on the following day, but the carrier was nowhere to be seen."[3]

During the same mission one of the bombs dropped by Lt. Walter Storp's crew hit HMS *Hood*, but it failed to explode and just bounced off the battlecruiser's deck. No other hits were reported.

Upon return to base, Gefr. Francke filed this rather succinct report: "Dive-bombing attack on aircraft carrier with two SC 500 bombs; first, near miss close alongside; second, possible hit forward quarter. Results not observed."[4]

The results of the first operation against the Royal Navy were eagerly anticipated by the Luftwaffe High Command. The events that followed the mission were very accurately described in Cajus Bekker's monograph on Luftwaffe's World War II history "Angriffshöhe 4000" (Attack Height 4000): "No sooner did Francke return to Westerland than a real circus began to swirl all around him. Only Oberst Siburg, CO of 'Löwengeschwader' (or Lions' Wing, KG 26 – author's note), remained skeptical: 'Did you actually see the ship sink?'

'No, Herr Oberst!'

'Well then, my dear boy', said Siburg with a broad smile, 'that simply means you didn't hit her.'

The common knowledge of naval warfare that signs of fire or even clouds of smoke did not necessarily indicate the sinking of a ship was foreign to Luftwaffe personnel.

In the meantime the telegraph cables at 10.Flegerdivision's HQ were white-hot. Brass at Oberkommando der Luftwaffe (Luftwaffe High Command) wanted to know why the Division's staff had not yet submitted a formal report on the sinking of a British aircraft carrier.

Major Martin Harlinghausen, the Division's chief of staff, responded: 'Because we have no knowledge of the sinking of that type of warship.' This was the truth, since all that Harlinghausen had was Francke's cautious report, which had been already forwarded to Berlin. However, the whole matter was beginning to live a life of its own. A major reconnaissance operation was mounted to establish what really happened to HMS *Ark Royal*. Finally, at 17.00, the first message arrived: 'Enemy force observed in grid square (...), two battleships and escort vessels, steaming westwards at high speed.'

The carrier was gone!

The most obvious option that the British force had simply split up did not occur to anyone in Berlin. *Ark Royal* could have left the area with the other half of the fleet, undetected by the German reconnaissance aircraft. Soon another order was radioed out to the patrol aircraft: 'Watch for oil spills!'

The CO of KG 30, Obstlt. Walter Loebel, preparing the crew briefing prior to a raid on Scapa Flow.

When a fortuitous oil slick was indeed found shortly thereafter, nobody seemed to be bothered by the fact that such 'evidence' was in abundance all over the North Sea. Under those circumstances, who could doubt that HMS *Ark Royal* and her sixty aircraft had been sent to the bottom?

Göring, Milch and Jeschonnek's only concern was whether they should wait for the official announcement from the British. However, the propaganda bandwagon was already spin-

An observer from Geschwader-Stab KG 30 is waving to the photographer after a return from a mission over Scapa Flow.

This crew of a Ju 88 A-1 of II./KG 30 were photograph after their return from a sortie over Scapa Flow.

ning its own version of the events: German air force sank the most advanced British aircraft carrier using one bomb. A single bomb! What a great feat!

When Hptm. Pohle arrived at Westerwald later that night (having been summoned there by Jeschonnek), Gefr. Francke ran up to him, forgetting completely proper military etiquette: 'Pohle, my man', he implored his CO, 'none of that stuff is true. Help me get out of this mess!'

Unfortunately, there was nothing Pohle could do to stop the tidal wave of events that followed. On the following day the communiqué of the Wehrmacht High Command reported the news of an air strike on the British naval force. The report read: 'In addition to the aircraft carrier that had been destroyed, numerous hits were scored on one of the battleships. Our crews suffered no losses.'

Göring also jumped on the propaganda band wagon. He wrote a personal commendation letter to Francke, promoted him immediately to the rank of Leutnant and awarded him the Iron Cross 1st and 2nd class.

At that point the British Admiralty went on the counteroffensive. The British released a dry statement confirming the return of the allegedly sunk *Ark Royal* to her home port. Photographs of the ship's arrival were made available to the press. (...)

Göring bore the grudge for months to come. On one occasion, when he bumped into

The officers of I./KG 30 debriefing a mission over Scapa Flow on March 16, 1940. From left to right: Oblt. Magnussen, Oblt. Philipps and Maj. Fritz Doensch.

Junkers Ju 88 A-1 being prepared for another operation against the Home Fleet.

Francke at Erprobungsstelle Rechlin, he said only half-jokingly: 'You still owe me an aircraft carrier!'[5]

On October 2, 1939 10. Fliegerdivision was re-named X. Fliegerkorps. Gen.Lt. Hans Ferdinand Geisler and Maj. Martin Harlinghausen remained in their posts as the CO and chief of staff. In addition to KG 26 and I./KG 30, the Corps was also in command of assets from LG 1 and KG 1.

A week later, on October 9, 1939, X. Fliegerkorps mounted a joint operation with Kriegsmarine against the British fleet operating in the North Sea. To lure the enemy out onto the open sea, Kriegsmarine sent out a force consisting of the battleship *Gneisenau*, light cruiser *Köln* and nine destroyers. In the morning the crews of three Dornier Do 18 flying boats from 3./Kü.Fl. Gr. 406 reported sightings of three British cruisers and two destroyers near Egersund. Shortly thereafter 127 Heinkel He 111 bombers from Stab, I. and II./KG 26, as well as II. and III./LG 1 launched from their bases in northern Germany, followed by 21 Ju 88s from I./KG 30. KG 1 remained on the ground as reserve.

Due to adverse weather and inaccurate reports of the position of the British force, most crews were forced to return to base without making contact with enemy vessels. Only a handful of He 111s from I./KG 26 and 4./KG 26 and Ju 88s from I./KG 30 managed to locate and engage their targets. The crews reported ten hits on the British cruisers, but in reality all bombs missed their marks. One of the Ju 88s from 3./KG 30 was hit by anti-aircraft fire and although the crew managed to nurse their machine almost within sight of the Danish coast, they eventually lost control and crashed into the sea. Two of the crew members perished in the crash, while the other two were picked up by search and rescue teams. This was the first Ju 88 combat loss of the war.

On October 16, 1939 the crew of an He 111 from Stab/KG 26 spotted a British battlecruiser, most likely HMS *Hood*, arriving at Firth of Forth. Cajus Becker provides an interesting account of the events that followed: "At 9.30 Jeschonek telephoned Hptm. Pohle in Westerland with operational tasking. He added: 'I also have a personal order from the Führer for you: if KG 30 arrives over Firth of Forth after *Hood* has already entered the shipyard, you are expressly forbidden from attacking her!'

After Pohle had acknowledged the orders, Jeschonek reiterated: 'I make you personally responsible for passing this on to every crew member. Führer does not wish to see any civilian casualties!'

Once again, the concern for minimizing collateral damage had utmost priority; neither the Germans, nor the British wanted to be the first to drop bombs on the enemy's territory. Warships were fair game, but only as long as they remained in the open sea. Once they were moored or docked in the shipyard any attack against them was expressly forbidden. Any use of military force remained subordinate to political goals as Berlin was still hoping for a swift end of war against Britain.

At 11.00 on October 16, 1939 fifteen Ju 88s from I./KG 30 took off from Westerland. At 14.15 the bombers reached Firth of Forth and followed it inland.

'We maintained a loose formation', reported Pohle, 'since according to 5. Abteilung reports (Department 5 of Luftwaffe's General Staff

– author's note) there were no Spitfires based in Scotland.'

To Pohle's dismay, it soon turned out the reports were less than accurate. British Fighter Command had two Spitfire squadrons (No. 602 and 603) operating from RAF Turnhouse near Edinburgh. Additionally, No. 607 Sqn (operating Hurricanes) arrived that morning at RAF Drem on the south shore of Firth of Forth.

The RAF squadrons' task was to intercept any incoming enemy bombers out over the sea, before they could reach the coast. One of the radar stations was dedicated to surveillance of the area, but it was down on that day due to a power failure. Unaware of the threat, the RAF fighters were not scrambled until the Ju 88s flew right over their bases at 4,000 m. The defenders lost precious minutes, which gave the German crews time to locate their targets.

Leading the loose formation, Pohle could see Edinburgh down below. For the first time during the war a Luftwaffe bomber formation appeared in the skies above Britain. Right in front of the incoming bombers was a large bridge across Firth of Forth. Just behind it, on the north shore, was the Royal Navy base and dockyards at Rosyth.

At that moment Pohle spotted what he thought had to be his primary target: she was much wider and longer than other ships. It had to be the *Hood*.

However, the warship was not at anchor, but already at the shipyard in a lock leading to the dry dock. It looked like she had arrived there just minutes earlier.

'She was a beautiful target', remembered Pohle, 'but we were not allowed to attack her...'

Nonetheless, Pohle put his Ju 88 in a dive. There was a multitude of cruisers and destroyers laying at anchor at Rosyth. Pohle looked for the largest target, which happened to be HMS *Southampton*. Anti-aircraft artillery responded immediately with very heavy barrage. While his Ju 88 was tossed around by shockwaves of exploding shells, Pohle pressed on in an 80° dive going straight for his target.

Then it happened. First there was a brief but extremely hard impact and then a deafen-

Ju 88 A-1 launching for a combat mission over Scapa Flow.

ing bang. Suddenly ice-cold rush of air hit the crew's faces.

The upper part of the Ju 88's canopy was gone just as the aircraft was screaming down in a wild dive at over 600 km/h.

Pohle had no idea if the bomber had received a direct hit, or if the damage was caused by overstressing the airframe. A similar thing happened to one of the bombers tested at Rechlin, so it could have been proof positive that the design was still not free of its teething problems.

Pohle kept his cool and continued the dive keeping the *Southampton* centered in his bomb sight. Just above 1,000 m he released the 500 kg bomb. They made it!

The bomb did hit the 9,100 ton ship's starboard side, but did not explode! The dud penetrated three decks and punched a hole in the ship's side before crashing into and sinking a small admiralty barge moored to the *Southampton*'s side.

The bombers crew had no time to follow the bomb's trajectory. No sooner did Pohle recover from the dive, the radio operator sitting behind him yelled: 'Spitfires at six o'clock!'.

'I was unable to perform any evasive maneuvers', remembers Pohle. 'The port engine was hit and almost immediately caught fire. All I could do was turn back towards the open sea hoping to reach the fishing boat *Hörnum*, which was ordered by the Kriegsmarine to remain close to the Scottish coast for just that kind of emergency.

However, the Spitfires were not about to give up. The 'wonder bomber' with its single MG 15 gun in the rear turret was practically defenseless against the British fighters.

In the second attack run Spitfire pilots peppered the bomber's cockpit with bullets, killing the rear gunner and radio operator. Flying over Port Seton East Lothian, Pohle dropped down low trying to shake off the attackers, but the RAF pilots hung on to his tail.

The third attack launched by the British fighter left the navigator severely wounded and put the bomber's starboard engine out of action.

This Junkers Ju 88 A-1, W.Nr. 299, 9K+HS of 8./KG 51 was photographed in the spring of 1940. The individual aircraft letter was red with a white outline.

Another shot of the same aircraft showing the wing's insignia under the cockpit.

The training version of the Ju 88 A lacked the defensive armament. Note the bomb ejectors under the wing's center section used for delivery of practice munitions in dive bombing training sorties.

This is how Pohle described those dramatic moments: 'That was the end. I spotted the trawler steaming north and hoped I'd be able to reach it gliding just over the waves. Then I passed out.'

The trawler's crew reached the sinking bomber within minutes and pulled Pohle, the sole survivor, out of the cockpit. Later on the still unconscious Hauptman was picked up by a British destroyer. When Pohle came to five days later he was in an infirmary at Port Edwards on the north shore of Firth of Forth."[6]

Credit for shooting down Pohl's Ju 88 (4D+AK) went to F/L Pinkerton from No. 602 Sqn. Another Ju 88 (4D+DH from 1./KG 30, flown by Oblt. Siegfried Storp) was downed by Spitfires from No. 603 Sqn. The bomber's rear gunner was killed, while the other three crew members were captured. There were also losses on the British side: P/O Gilroye from No. 603 Sqn. was killed when his Spitfire (s/n L1048) was shot down by one of the Ju 88 gunners.

Lt. Horst von Riesen's attack was much more successful. Two bombs that he dropped on the destroyer HMS *Mohawk* exploded alongside the ship causing heavy damage and killing 13 of the crew, including the ship's commander. Another 30 crew members were wounded in the attack. The destroyer was so heavily damaged that for the next two months, until January 14, 1939, she remained docked in the shipyard for repairs.

After the attack Lt. von Riesen's Ju 88 was jumped by three Spitfires. This is how he later remembered the fight: "I was sure that was the end of it. Their machine guns fired nonstop from all directions and it seemed like the Spitfires were taking turns getting on my tail. Perhaps they were surprised by my speed, but that's what eventually saved me – I was doing well over 400 km/h."[7]

Junkers Ju 88 A-1, VA+EH, "yellow 10" during a test flight at factory airfield prior to delivery to a combat unit.

This Junkers Ju 88 A-1, 9K+BP was operated by 6./KG 51 "Edelweiß".

Von Riesen was heading out towards the open sea. In front of him and down below he saw what looked like drops of rain hitting the water. He quickly realized those were rounds that were meant for him. In the early days of the war it was common practice for RAF armorers to set up the eight machine guns carried by their fighters for maximum coverage at long distances. With that setting the pilots stood a very decent chance of scoring good hits, but at the same time no chance at all for proper aiming. Later on the practice was abandoned in favor of setting the guns' convergence at a point some 230 m ahead of the fighter's nose. Had the guns been set up like that, von Riesen would not have had a chance to survive. After a mad chase that lasted about twenty minutes, the Spitfire pilots realized they were too far out from their home base and were forced to break off, but not before they shot up the cooling system of the Ju 88's starboard engine. Von Riesen had to shut down the crippled engine, which brought his speed down to 180 km/h, just barely above the bomber's stalling speed. The Ju 88 was notoriously difficult to fly on the best of days and even more so when the pilot had to struggle with asymmetric thrust. It was also very difficult to maintain height with one engine inoperative, but somehow von Riesen managed to keep the bomber flying; he even gained a bit of altitude as the fuel was being burned. After a grueling four hour flight von Riesen brought his Ju 88 for an emergency landing at Westerland. The bomber's battle damage was estimated at 45%.[8]

Another operator of the Junkers Ju 88 A was LG 1. The unit's badge can be seen under the cockpit.

Junkers Ju 88 A-1, 9K+A? of KG 51 "Edelweiß" photographed in April 1940.

Another photograph of the same aircraft shows the early type of the B-Stand mounting a single flexible MG 15 7.92 mm machine gun. Two additional MG 15s are mounted in ports in the canopy's side windows.

On the same day Hptm. Fritz Doench was appointed the new CO of I./KG 30. A day later he led a four-ship flight of Ju 88s on a raid against Scapa Flow, the Royal Navy's main naval base in the Orkney Islands. The Junkers bombers were accompanied by 13 Heinkel He 111s from I./KG 26. When the German force arrived over Scapa Flow they found the anchorage almost deserted as the British fleet had moved to Loch Ewe on Scotland's west coast following an attack by the U 47 and the sinking of HMS *Royal Oak*. The only ship still there was an old decommissioned battleship HMS *Iron Duke*, which had been converted into a base ship and a floating anti-aircraft battery. One of the bombs dropped by the Ju 88s exploded alongside the battleship causing damage below the waterline. In order to save the ship her crew ran the *Iron Duke* aground. In the meantime, one of the Ju 88s from 2./KG 30 was hit by the battleship's anti-aircraft fire. The bomber, flown by Oblt. Walter Fläming, crashed on the island of Hoy killing all but one of the crew.

The Ju 88 crews from I./KG 30 flew their next combat mission on October 21, 1939. At 12.35 the bombers took off from Westerland to attack a convoy, which had been detected

near Flamborough by a German reconnaissance aircraft. However, the British anti-aircraft fire kept the raiders at bay and none of the bombs reached their intended targets.

On Sunday, November 13, 1939 the Ju 88s appeared for the first time over the Shetlands. The raid mounted by 13 bombers from I./KG 30 was very interestingly described by one of the German bomber crews: "The weather was great with unlimited visibility. We were cruising at around 500 m thinking that nobody would be able to hide from us in such wonderful conditions. However, as is commonly known, soldiers should refrain from thinking. Ninety minutes later things looked entirely different. The visibility was all but gone, squalls made it difficult to maintain height and heading and rain was lashing us from all directions. The sea below was simply boiling. It looked like Neptune was having a really bad day. We didn't mind any of that. Not at all! All we wanted was to finally drop our bombs on the enemy. In the meantime, the weather was getting worse by the minute. Soon it was so bad that even the toughest of us, with lots of combat time in stormy weather were beginning to have doubts. Still we hoped all that bad stuff would soon clear and we'd be right over the target with the enemy in sight.

Junkers Ju 88 A-1, W.Nr. 3134, 9K+EL from 3./KG 51 photographed at Memmingen.

Another photograph of the same machine. The aircraft's individual letter "E" was yellow, while the serial was painted white.

Junkers Ju 88 A-1, W.Nr. 8020 from II./KG 30.

And that's exactly what happened. Suddenly the Shetlands came into view with their high cliffs lashed by waves. I'll be damned! It looked like the entire British fleet was moored down there! We quickly turned towards what we thought were British ships, only to find out they were in fact just large rocks. In almost zero visibility that kind of mistake could be forgiven. We could all swear we saw enemy ships down there. Oh well, damn shame! Here and there we spotted

Junkers Ju 88 A-1 of KG 30 (in the upper right-hand corner) flying over Stavanger-Sola airfield captured by German paratroopers.

A Ju 88 A-1 pilot in his office. On the right is the A-Stand MG 15 machine gun, while the stowed Revi C 12/C gun sight is visible on the left.

small fishing boats, but apart from them we could see nothing except green meadows and herds of sheep. So where is that proud Home Fleet? In what dark hole are they cowardly hiding? Over the Shetlands we split up. We left the HQ flight and 1st squadron to comb the area on their own. The weather forced us to drop closer and closer to the ground: 300, 200 and finally 100 meters. And we still couldn't find a target for our bombs. The bombardiers were cursing like mad since they knew time was running out and soon they'd be forced to jettison the bombs. Still hopeful, we made one more wide turn, but found nothing. Time was up, we had to turn back to base. That was a damn shame, we thought, but what could we do? We had no choice but to head home.

But then lady luck smiled on us, ever so slightly, as we spotted a large, four-engine flying boat moored close to the shore. Before the

The observer station in the Ju 88 A-1.

seven crew members could realize what was going on we were on top of them. We dropped to 30 meters and the entire squadron began to circle the flying boat, just like horses running around the Renz circus arena.[9] It took just half a magazine of ammunition fired from one of the guns and that was that: fire, a pillar of smoke and England was worse off by one flying boat. A million marks went to the bottom."[10]

On December 22, 1939 a two-ship of Ju 88s from I./KG 30 was flying a long-range reconnaissance mission over Firth of Clyde at 4,800 m when they were attacked at 10.29 by Spitfires from No. 602 Sqn flown by F/L J.D. Urie, F/O C.H. Maclean and F/O Strong. The British pilots mistook the Junkers bombers for Heinkel He 115 flying boats carrying naval mines under their wings. Despite several attacks, the RAF fighters failed to shoot down the bombers, which managed to return to base safely. One of the machines sustained battle damage estimated at 50%. One of the Junkers crew members provided this account of the mission: "The North Sea was awash with moonlight during the first hour of the sortie. As on other occasions over the last several weeks we were unable to spot any shipping. The sea was empty as far as the eye could see. It wasn't until we approached the coast that we could finally make out several brightly lit freighters. As we continued the mission the weather began to deteriorate, until the skies were completely overcast. After we had reached the east coast of Britain we turned straight for the Firth of Clyde. We continued over land at just 50 meters. It was an eerie feeling to be flying that low over Britain. We could see people in the streets. We saw passengers getting on a train as we flew over a train station. Not once did we notice any signs of air raid alarm, although we knew the English must have seen us. It looked like their early warning system was just not as good as ours. I could simply not imagine an enemy reconnaissance aircraft flying over the Reich at 300 meters completely unnoticed and unmolested by flak! During our entire flight not once were we fired upon by the flak batteries.

When we were approaching the east coast again and getting ready to set course for home, a Spitfire appeared out of the blue and opened up on us. In the meantime, the clouds thinned out a bit and we were flying some 100

Member of ground personnel from KG 30 preparing an SC 500 to be hung on one of the underwing stations of a Ju 88 A.

ETC 500/IX d bomb ejectors.

meters above them. The Englishman broke left and came in to attack us from above. During previous missions we were often fired upon by anti-aircraft guns, but never by fighters. We kept our cool and decided a few machine gun muzzles pointing at us was not the worst that could happen. We too had our own sprinklers! But the Englishman was a gutsy type. During his first two passes our perfectly composed radio operator (who, by the way, was on his first combat mission) fired two full magazines straight into his cockpit. The Englishman broke off the attack and we used the time to head for the clouds. We knew the English pilot must have thought we had gone down. As we later found out three patrol boats were dispatched to pick us up. We too thought the Englishman was no more since we could not find a good explanation for his sudden dive from the height of just 100 meters.

When we broke out of the clouds we were over the North Sea and the English fighters were nowhere to be seen. Now was the time to take stock of the situation and assess any damage to our aircraft, at least the bits we could do in flight. The airspeed indicator was inoperative, which complicated matters during our return leg. What's worse, there was a fuel leak from the left wing. Damn it, we thought, let's just hope we'll have enough to make it back home.

In the meantime we found ourselves in a bit of trouble. The sea below was boiling. Huge swells rose upwards with white froth on top stirred by heavy winds. It's such a terrible feeling not knowing what might happen next. We spent over three hours flying through heavy rain and snow praying not to crash into the abyss below.

But we made it. A patrol boat welcomed us as we approached the home plate and soon we checked in with the tower. We managed quite a smooth landing, considering one of the tires was shot up, which in fact we didn't know at that time. The rounds from Spitfire's guns not only hit the portside main landing gear tire, but also damaged gear extension mechanism and pushrods actuating the flaps."[11]

December 1939 saw the establishment of II./KG 30, whose personnel was drawn from Lehgruppe 88. KG 30's strength grew further when III./KG 30 became operational in January 1940. Both units and the wing's HQ (Geschwader-

In a few moments this Ju 88 A-1 will begin its take-off roll.

stab) operated from Barth air base on the Baltic coast. Obstlt. Walter Loebel, former CO of KG 26, became the wing's new commander.

In the morning of January 1, 1940 six Ju 88s from I./KG 30 launched an operation against British shipping. When four of the crews were dropping bombs on ships between Scapa Flow and the Shetlands, the other two came under attack by a pair of Hudsons from No. 220 Sqn. Before long two Gladiator fighters joined the battle. Following the fight single Ju 88 kills were reported by F/O T.W. Gillen and F/O R.A. Winter. The crew of one of the Hudsons (individual a/c identifier "L", flown by P/O W.F. Carey) also claimed a single air-to-air victory. In fact, only one Ju 88 went down in the engagement and crashed into the sea. The other Hudson ("T", s/n N7232) was heavily damaged by the Junkers gunners and crashed some 32 km from Saint Abb's Head killing F/L T.H. Clark and his crew.

On January 3, 1940 X. Fliegerkorps mounted another anti-shipping operation, this time targeting the area between the Orkneys and Britain's east coast. The first wave consisted of 14 Heinkel He 111s from II./KG 26 and a pair of Ju 88s from I./KG 30. Following them were ten He 111s from I./KG 26. The incoming bombers were soon picked up by British radar and a scramble signal was sent to the fighter crews. When the RAF fighters engaged incoming He 111s, the two Ju 88s attacked British minesweepers. The Royal Navy ships responded with accurate anti-aircraft fire knocking down one of the attacking Ju 88s. The bomber crashed in Morray Firth, but its crew survived and was later picked up by the British ships. However, the

Junkers Ju 88 A-1 of KG 30 photographed during the Norwegian campaign in the spring of 1940.

loss of the Ju 88 was not in vain as the Germans scored their first victory sinking the British minesweeper HMS *Sphinx*[12]. The ship was hit by a single bomb, which penetrated the deck and killed five of her crew, including the CO, Cdr. John Robert Newton Taylor. HMS *Boreas* picked up 46 survivors, but 49 lives were lost when the *Sphinx* sank.

On February 9,1940 eight Ju 88s from I./KG 30 and nine He 111s from II./KG 26 launched an attack on a convoy steaming along Britain's east coast. The Junkers crews dropped bombs on a flotilla of minesweepers sinking two of them (HMS *Robert Bowed* and HMS *Fort Royal*). Four officers and eighteen sailors were killed in the attack. British anti-aircraft gunners hit one of the Ju 88s, whose crew attempted to nurse their machine back home. However, the bomber never made it to the home plate and crashed into the sea north west of Sylt. The crew, led by Fw. Friedrich Pfeiffer, was listed as missing.

A month later, on March 9, 1940, a three-ship of Ju 88s from I./KG 30 arrived in the skies over Scapa Flow to conduct reconnaissance of the Home Fleet anchorage. Unfortunately, the German crews were unable to spot any surface vessels due to fog covering the waters of the bay. Moments after the bombers had set course for home, four Hurricanes from No. 111 Sqn. launched from their base and gave chase to the German formation. F/O R.G. Dutton was the first one to catch up with the Ju 88s, which by then were flying some 65 km from the Orkneys. He promptly downed one of the bombers, but misidentified his victim as a Heinkel He 111. The Junkers crew, commanded by Oblt. Frithjof von Sicharthoff, were killed in action.

In the evening of March 16, 1940 Scapa flow became the target of 21 He 111s from I./KG 26 and 18 Ju 88s from I/KG 30. Later one of the Junkers crew members, Oblt. Hans-Werner Magnus-

One of the Royal Navy's auxiliary ships under attack by Juners Ju 88s of KG 30.

This Ju 88 A-1 from KG 30 is being loaded with bombs. Note the yellow tips of propeller spinners.

The crew of this Ju 88 A-1 from KG 30 were photographed prior to another combat sortie over Norway.

sen, offered this account of the action: “We heard the call to commence the bomb run and went through the drill as we had done many times before, both in peacetime training and in combat. The bombardier lay prone in the nose section of the cockpit and quietly gave me, his pilot, steering commands: a bit to the right, slightly left – there, that’s good! After a few seconds the flight engineer reported flak off to our right, but there was no time for nuisance of that kind when you were on your bomb run. Finally came the much awaited command from the bombardier: attention – zero! The aircraft shuddered, which was a tell-tale sign that we had just rid ourselves of what was at the same time our greatest burden and favorite pet. Now all our thoughts were focused on that heavy package that we had just sent on its way down with our heartfelt wishes. A few seconds had passed before I got the confirmation of our success. The first thing I heard was the rear gunner raising the squadron’s battle cry. Then came his happy voice screaming over the intercom: ‘Direct hit, Herr Oberleutnant! The target’s on fire!’”[13]

Ju 88 A of KG 30 coming in to land at a Norwegian airfield.

Several Ju 88s crews were tasked with hitting airfields around Scapa Flow to prevent the RAF fighters from intercepting the strike force before it reached the target area. The bombers were led by Oblt. Wolfram Philipps from I./KG 30: "To our great delight we saw several huge explosions among the muzzle flashes of flak guns, which couldn't have been anything else but the bursts of heavy bombs dropped by our comrades. That was the first thing I noticed before I commenced my bomb run.

In the falling darkness we were invisible against the dark sky, while we could clearly see the Orkneys and our target: Earth House air base with its hangars and operating area. A pair of English fighters tried to attack my bomber from below, but they failed and I was able to release the entire bomb load over the airfield as planned. We could clearly see our bombs deto-

The British steam ship *Royal Crown* destroyed by German bombs in Norway.

Ju 88 A-1 of KG 30 parked at Trondheim/Vaerne.

The A-Stand in the Junkers Ju 88 C-2 destroyer. A Revi C 12/C reflector sight is visible on the right, in front of the pilot's seat.

nating one by one in a neat row, with columns of smoke rising up above the flashes of explosions. Within seconds there was lots of bright fire, which was a tell-tale sign our mission was a complete success.

When we were still over the target area we spotted English fighters trying to take a stab at us coming in from below. However, they probably didn't launch until they saw us right over their own airfield, so they never had a chance to climb to our altitude, or to prevent us from dropping the bombs. In the meantime it got so dark that enemy fighters lost whatever little chance they still had to intercept and attack us, so our formation pressed on and attacked all three airfields that had been our designated targets without any opposition from English fighters.

As I was dropping the bombs English flak gunners tried to take a poke at me, but in the end none of our machines was hit – nothing, not even a scratch."[14]

In total, the Germans dropped around 100 bombs and claimed hits on three battleships and a battlecruiser. In reality the battlecruiser HMS *Norfolk*[15] was damaged following a hit below the waterline, while three other bombs exploded alongside an old battleship (then serving as a base ship) HMS *Iron Duke*, which caused very little damage. Six of the *Norfolk* crew died in the attack and the ship herself spent the next three months undergoing repairs. That night also saw the first British civilian casualty of the war: 27-year-old James Isbister of Bridge of Waithe, a village on Loch Stennes, was killed in his own house. Seven additional civilians were wounded in the air raid.

On the return leg back to base one of the crews from 3./KG 30 (commanded by Uffz. Werner Mattner) lost their bearings over the North

Sea and were forced to make an emergency landing on the Danish island of Lolland, near Cöbelen, where they were promptly interned.

In the morning of March 17, 1940 six He 111s from II./KG 26 followed by a four-ship of Ju 88s from 6./KG 30 were spotted just off the coast of Scotland. Four Spitfires from No. 603 Sqn. launched from their base at 08.45 to intercept the bombers. One of the Ju 88s came under attack twice: first at 09.15 (F/L G. Denholm), then at 09.20 (F/O Haig). The bomber sustained minor damage, while the radio operator was wounded in the attack.

The mission was an operational debut for the crews of II./KG 30, which had just reached full combat-ready status. During the unit's second operation, launched on March 29, 1940, nine Ju 88s attacked a coastal convoy south-east of Sunderland. One of the aircraft from 6./KG 30 (4D+AP, flown by the squadron's CO Oblt. Rudolf Quadt) was hit by anti-aircraft fire and crashed into the sea killing the entire crew.

In the evening of April 2, 1940 nine Ju 88s from Stab and II./KG 30 mounted an attack on Scapa Flow. The mission was a failure, although one British soldier and two civilians were killed by shrapnel. One of the German bombers from Stab KG 30 was damaged by anti-aircraft fire and crashed on the way back to base, having flown some 200 km over the North Sea. The crew, commanded by Oblt. Bülow, perished in the crash. Commander of 4./KG 30 and his crew were killed when their bomber crashed in bad weather (fog and low clouds) some 25 km east of Bremerhaven.

On the following day, April 3, 1940 four Ju 88s from 5./KG 30 and 21 He 111s from I./KG 26 attacked British convoy ON 24 off the coast

The British battleship HMS *Warspite* being attacked in Vestfjord by Ju 88s from KG 30.

Two SC 500 bombs exploding just of the stern of HMS *Warspite*.

A wreck of a British trawler beached in a Norwegian fjord after a Ju 88 attack.

This Ju 88 A of 1./KG 30 bears the scars of AAA shrapnel.

of Norway, some 100 km north-west of Bergen. The convoy was escorted by a four-engine Short Sunerland flying boat (a/c identifier "F", s/n N9046, flown by F/L Frank Phillips). Two of the Ju 88s attacked the Sunderland, coming in from astern. The flying boat's rear gunner, Cpl. W.G. Lille waited until the first of the attacking aircraft came within 100 m, then opened fire with his quad 7.62 mm Brownings. The first long burst fired by Lille found its mark and the Ju 88 nose-dived into the sea, killing Oblt. Karl Overweg (CO of 5./KG 30) and his crew. In the meantime the Sunderland's co-pilot, P/O Armittage, coordinated the efforts of waist and top turret gunners. Within minutes the second Ju 88 was limping back towards the Norwegian coast trailing black smoke coming out of one of its engines. Uffz. Willi Erkense managed to crash-land his damaged machine near Stavanger. The aircraft was completely destroyed in post-crash fire, but Erkense and his crew survived and were subsequently interned.

Operation Weserübung

One of the units that took part in operations against Denmark and Norway was KG 30 operating Ju 88 bombers. The wing was part

Lt. Werner Baumbach in the cockpit of his Junkers Ju 88 A-1 of 5./KG 30.

The French cruiser *Emile Bertin* hit by a bomb dropped by Lt. Werner Baumbach from 5./KG 30 on April 19, 1940.

Lt. Werner Baumbach from 5./KG 30 was awarded the Knight's Cross on May 8, 1940.

of X. Fliegerkorps and comprised the following units (see table on the right).

Unit	Airfield	Commander	Number of aircraft	Aircraft types
Stab/KG 30	Westerland	Obstlt. Walter Loebel	1 1	He 111 P Ju 88 A
I./KG 30 (without 1. Staffel)	Westerland	Hptm. Fritz Doench	25	Ju 88 A
II./KG 30	Westerland	Hptm. Claus Kinkelbein	30	Ju 88 A
III./KG 30	Barth	Hptm. Siegfried Mahrenholtz	13	Ju 88 A
Zerstörerstaffel KG 30	Westerland	Oblt. Herbert Bönsch	6	Ju 88 C-2

Following the landings in Norway, the Germans expected a swift response from British naval forces and planned to use the Ju 88s from KG 30 as the main strike force against the Royal Navy assets.

German invasion got underway on April 9, 1940 and, as expected, the Home Fleet's response was instantaneous. A force of Royal Navy ships that were at sea at that time (four battleships, three battlecruisers and six light cruisers, escorted by a flotilla of fourteen destroyers) immediately set course for Bergen.

The British warships were detected by Luftwaffe reconnaissance aircraft and soon the X. Fliegerkorps strike aircraft were scrambled to attack the approaching enemy. Between 13.50 and 16.00 the Germans launched 41 He 111s from I. and II./KG 26, 22 Ju 87 Rs from I./St.G 1 and 47 Ju 88s belonging to KG 30 to intercept the British force. The mission turned out to be beyond the Ju 87 range, forcing all Stuka crews to abort and return to base.

The attack began at 15.30 when the first bombs were dropped by the first wave of KG 30 crews. HMS *Southampton* sustained minor dam-

Ju 88 C-1, W.Nr. 0142, 4D+MH, Stammkenzeichen VB+WC was shot down on June 7, 1940 while attacking ground targets at Narvik. The crew were forced to crash land the bomber near Bjornefjell.

Rotterdam-Waalhaven airfield was one of the targets of a dawn attack launched on May 10, 1940 by Ju 88s from KG 30.

age caused by shrapnel from the German bombs that exploded alongside the ship and put the cruiser's anti-aircraft artillery fire control center out of action. One of the destroyers (HMS *Gurkha*[16]) had less luck: she was severely damaged when a bomb exploded just off her stern punching a 2 m^2 hole in the hull just below the main deck. The explosion set several aft compartments on fire and flooded the engine room. The ship sat dead in the water until 20.45 when she capsized and sank killing 5 officers and 11 enlisted men. 194 survivors were picked up by the cruiser HMS *Aurora*, while destroyer HMS *Mashona* rescued five of *Gurkha*'s crew members.

Minutes later the remaining Ju 88s arrived over the target area, followed by He 111s from KG 26. Following the attack the Luftwaffe crews claimed good hits on three Royal Navy battleships. In reality one of the 500 kg bombs did hit the deck of HMS *Rodney*, but failed to penetrate the ship's 150 mm armored deck. Ten of the battleship's crew were wounded in the attack. Battlecruiser HMS *Devonshire* suffered only minor shrapnel damage when German bombs exploded alongside the ship. The damage had no impact on the battleship's ability to continue operations. The bomb that exploded alongside light cruiser HMS *Glasgow* killed two of her crew and wounded five more. The ship's main artillery 'A" turret was also briefly put out of action, but all damage was repaired over the next couple of days when the cruiser was still at sea.

The second wave of German bombers met significantly stronger resistance from the ships' anti-aircraft batteries than the lead element of the strike. Four of the attacking Ju 88s were shot down with a loss of all crews. Among them was the Ju 88 A-2 from Stab III./KG 30, which was flown by the *Gruppe*'s CO Hptm. Siegfried Mahrenholtz. The other three machines lost that day

The Knight's Cross recipient, Lt. Werner Baumbach from 5./KG 30 flying a combat sortie in May 1940.

were the Ju 88 A-1 from 4./KG 30 (pilot: Fw. Hans Sültmann) and two Ju 88 A-2s from 3./KG flown by Ofw. Ernst Fahr and Uffz. Herbert Lipke.

Although the attack inflicted only limited damage on the British force, its psychological effect was significant. Following the strike, Home Fleet's commander-in-chief, Admiral Forbes, ordered his ships to withdraw beyond the range of the Luftwaffe bombers, especially in the area off the southern coast of Norway. Forbe's decision was partially dictated by the fact that during the attack the Royal Navy warships used over 40 percent of their supply of 102 mm antiaircraft munitions. It also transpired that the AAA fire control systems had problems dealing with fast flying German dive bombers. 12.7 mm machine guns were completely ineffective, while the 40 mm cannons were not available in adequate numbers. Additionally, most of the destroyers could not properly employ their antiaircraft weapons due to rough sea conditions. When the commanding officer of HMS *Gurkha* ordered his ship to slow down to 15 knots in order to improve stability of his AA guns, he quickly understood his mistake when the ship was hit by a well-placed series of bombs.

On the following day, April 10, 1940 19 He 111s from I./KG 26 and 19 Ju 88s from I. and II./

Junkers Ju 88 A-1 z 8./KG 4 "General Wever", 5J+CS. The letter "C" was red with a white outline. The aircraft was photographed during fighting in France.

A reconnaissance Junkers Ju 88 from 1.(F)/123.

KG 30 launched an evening sortie against fuel plants at Scapa Flow. Three of the Heinkels were lost in a fight against the RAF fighters, while the British anti-aircraft gunners scored hits on two of the Ju 88s. The first machine to receive hits was flown by Ofw. Walter Brünn from 2./KG 30. The crew reported the loss of one of the engines before all radio contact was lost. Another Ju 88 whose crew went missing following the raid belonged to 4./KG 30 and was commanded by Lt. Hans Hohendahl.

This photograph of the Junkers Ju 88 A-1 from KG 30 shows an additional flexible MG 15 machine gun mounted in the aircraft's glazed noose.

April 11, 1940 saw the arrival of autonomous (Z)./KG 30 at Sola airfield in Norway. On the following day three of the destroyer squadron's Ju 88 Cs and ten Bf 110s from ZG 76 attacked a formation of Wellingtons from No. 149 Sqn. Two of the Ju 88 crews reported air-to-air kills in the battle: at 16.45 Ofw. Martin Jeschke shot down Wellington P9266 flown by Sgt. G.E. Good, while five minutes later Uffz. Peter Lauffs destroyed Wellington P9246 piloted by Sgt. H.J. Wheeler. The Wellington gunners hit one of the attacking Ju 88s (flown by Uffz. Erwin Maus), which went down just south of Sola killing the entire crew.

On the following day, April 13, 1940 Peter Lauffs from (Z)./KG 30 stumbled upon a single Hudson some 10 km east of Stavanger. Lauffs reported shooting down the British aircraft, but his claim was rejected due to lack of independent witnesses.

Over the next several days the crews of KG 30 performed a series of maritime reconnaissance missions west of Aalesund. During one of the patrols, on April 15, 1940 three Ju 88s from 3./KG 30 jumped a Short Sunderland flying boat. Despite heavy damage and wounds suffered by one of the crew members, the British aircraft managed to return safely to Scapa Flow. On the following day Uffz. Peter Lauff and Oblt. Herbert Bönsch from (Z)./KG 30 each claimed a Blackburn Roc kill over Narvik.

Just before dawn on April 17, 1940 battlecruiser HMS *Suffolk*[17] approached Stavanger and shelled the German air base located there. Just before HMS *Suffolk* began the bombardment, a single Hudson from No. 233 Sqn (flown by P/O Edwards) arrived in the area to illuminate the target with flares and incendiary bombs. One of the flares landed on a Ju 52/3m parked at the airfield and set the machine on fire. The Junkers (belonging to Stab KG 30) was completely destroyed. The battlecruiser also launched her embarked Walrus flying boats, whose crews were to direct the ship's fire. The bombardment from a distance of 18,000 m lasted 80 minutes. The *Suffolk*'s crew fired 202 203 mm shells, but they hit a flying boat base rather than the airfield. Four Heinkel He 115 Cs from 1./Kü.Fl.Gr. 106 and a single He 59 D from K.Gr.z.b.V. 108 were destroyed in the attack.

In the pre-dawn twilight, Ofw. Martin Jeschke from (Z)./KG 30 launched an unsuccessful attack against the Hudson and then shot down one of the two Walrus flying boats. After his return to base the German pilot reported shooting down a Blenheim (sic!).

At about the same time HMS *Suffolk* ceased fire and turned northwards accompanied by her destroyer escort to the area where a force of German destroyers had been spotted by the crew of a British patrol aircraft.

In the meantime ten Heinkel He 111 bombers from I./KG 26 launched from Sola at 08.15

Ju 88 A-1 undergoing field maintenance.

Junkers Ju 88 A-1 of II./KG 30 ready for another sortie. Note the late-type B-Stand mounting a pair of flexible MG 15 machine guns.

Junkers Ju 88 A-1 from 2./KG 51 photographed on a soggy airfield in the spring of 1940.

Junkers Ju 88 A-1 of KG 51 photographed during fighting in Western Europe in May 1940.

Ju 88 A-1s from 1.(F)/123. The machine in the foreground is 4U+DH.

and attacked the battlecruiser and two of the escort destroyers. HMS *Kipling* sustained minor damage when two SC 500 bombs exploded alongside the ship. However, the British force commander took the Luftwaffe threat seriously and ordered his ships to return to Scapa Flow.

Between 10.00 and 10.17 a formation of 28 Ju 88s from KG 30 launched from their base and gave chase to the British ships. Twelve of the bombers located the retreating *Suffolk*, but only ten crews attacked the ship. Two bombers were forced to abort due to mechanical malfunctions. The Junkers crews attacked individually or in three-ship formations, diving in on their target from altitudes of 4,000 to 5,000 m. At 10.37 one of the bombers commenced a solo attack from 3,300 m. Diving in at 65° to 70° angle the crew released a single SD 1000 bomb at 1,400 m. The bomb hit the battlecruiser just forward of the main artillery "X" turret, penetrated the main deck and exploded between the aft engine room and 203 mm ammunition magazine supplying "X" turret. The explosion caused a cordite fire and put both aft main artillery gun turrets out of action. Tons of seawater flooded the ship's aft engine room, which immediately reduced *Suffolk*'s speed to just 18 knots. 32 crew were killed in the attack and another 37 suffered wounds.

At 13.25 another lone Ju 88 attacked the ship dropping three heavy bombs, which exploded some six meters off the battlecruiser's side, near the aft main artillery turrets. The shrapnel added to the already extensive damage and more leaks and flooding caused the ship's stern to almost completely disappear under the surface.

Several minutes later the heavily damaged ship found herself in range of friendly aircraft, whose crews arrived on scene and chased away a formation of eleven He 111s from KG 4. The Germans managed to drop only a single SC 250

One of the first Ju 88 A-1s lost by KG 51 during fighting over France.

The crew of a Ju 88 A-1 from 3./KG 51 are preparing for another mission at a French airfield.

bomb, which exploded near the ship's stern. On the following day HMS *Suffolk* arrived at Scapa Flow where she remained until December 1940 undergoing repairs.

On April 18, 1940 six Ju 88s from I./KG 30 mounted a reconnaissance operation in the area of Namsos. Despite adverse weather conditions, one of the crews spotted a British convoy consisting of four warships and several merchant vessels. The convoy's anti-aircraft artillery opened fire on the Germans and hit one of the Ju 88 As from 3./KG 30 (4D+BL),forcing its crew to make an emergency landing near Namsos, in which Lt. Hermann Peter lost his life. The remaining crew members (Ofw. Eugen Drexler, Fw. Rudolf Napperell and Uffz. Günther Burmeister) were captured and sent to a POW camp in England aboard one of the British ships.

In the morning of the following day eight Ju 88 As from KG 30 arrived over Narvik having previously refueled at Trondheim. One of the crews spotted the British battleship HMS *Warspite* lying in Vestfjord and attacked the ship with two SC 500 bombs, which missed the target. At the same time a three-ship of Ju 88s stumbled upon a convoy steaming past the mouth of Namsfjord escorted by destroyers and a French cruiser *Emile Bertin*. Lt. Werner Baumbach from 5./KG 30 remembers: "There was a waterway leading towards Aandalsnes and out to the North Sea. As soon as we went feet wet we hit a horrible snowstorm, which was so thick that we could only maintain formation by keeping an eye on the other machines' shadows cast against the falling snow. We flew on making a series of course corrections.

After a while we climbed and put on our oxygen masks. Before long the cloud cover thinned out and we could finally get a fix on our position looking down on the distant ground crisscrossed with fjords. Off to the distance was snow-covered Norway with its glaciers. It would be another half an hour before we reached our target – a force of enemy warships that had been spotted by our patrol aircraft.

Down below a spring storm stirred up huge waves the size of houses. It felt like we were just a toy in the hands of the raging elements.

The pilot of a Ju 88 A-1 from KG 51 signals his ground crews before another operational sortie.

As far as we could see there was only water, sky, clouds and... more water.

We began to prepare for the attack: we checked and double-checked the bomb ejectors and charged the machine guns. We were ready. The last few minutes seemed to drag on forever as the cockpit was filled with the atmosphere of tension and anticipation. I kept nagging our navigator who sat there hunched over his charts: 'Menz, are we over the target yet? It has to be somewhere around here!'. Indeed, before long we did find ourselves over the enemy ships. Almost simultaneously all crew members identified the target and began screaming: 'Battleship!'. Our gunner added: 'A big warship just below us and huge transports all around!'

From that moment on we dared not lose sight of our pray. At first I also thought we were dealing with a battleship, because she looked enormous in the tight confines of the fjord. However, when we got closer I could clearly identify characteristic features and shape of a cruiser. I couldn't commence my bomb run immediately, since at that very moment clouds gathered above the target. I had no choice but to turn around and make another approach.

Then the cruiser and transports opened up on us. It looked like the cruiser's gunners were just setting up barrage fire, since the sky above the ship was full of white puffs of exploding flak shells. During my second approach I saw my wingmen find a gap in the cloud cover and com-

A pair of Ju 88 A-1s from KG 51.

Ju 88 A pilot and observer during a combat sortie.

mence their dives. I saw the bombs going off just forward of the ship's bow. It also appeared that the flak fire was not as heavy as just a few minutes before. I was really intent on dropping my bombs on the second run, since the weather was getting worse by the minute. I just didn't want that cruiser to escape. I caressed the pickle button with my left thumb and adjusted my dive bombsight. I could see the sight's bright circle right in front of my eyes. The navigator, gunner and radio operator were all clutching their machine guns.

We were approaching our target. The cruiser was zigzagging trying to evade our attack. I found a small gap in the overcast and commenced the bomb run. As soon as we broke out I caught the ship in my sights and kept her there. We must hit her!

Bombs away! Then a sudden relief when the aircraft shuddered confirming good release of weapons. As I was recovering from the dive, the gunner shouted: 'A good hit amidships, on the starboard side!'

Our machine guns raked the transports trying to silence their flak guns. We all felt like flying through the eye of a storm. The throttles had already been shoved all the way forward as the snow-capped mountains loomed ever bigger in front of us. At the last possible moment I turned sharply to avoid hitting one of the peaks. In the meantime the gunner was furiously snapping away taking pictures of our target. The ship was obscured by smoke and fire of the explosions and it was clear she was beginning to sink. We made one more orbit over the target and then climbed up into the overcast.

Turning back to base we radioed the message of the successful attack. When we arrived over the field I saw the wing commander's car heading towards our dispersal area. As soon as we landed and taxied in he was there waiting for us to receive my formal report."[18]

However, the French cruiser did not sink. Heavily damaged, she limped back to Scapa Flow and then back to Brest. For his successful attack on the cruiser *Emile Bertin* Lt. Werner Baumbach was awarded the Iron Cross 1st Class on May 5, 1940. Three days later, on May 8, 1940 he became the first Luftwaffe bomber pilot to receive the Knight's Cross.

On April 20, 1940 a four-ship of Ju 88s of 2./KG 30 launched an unsuccessful attack on the cruiser HMS *Carlisle* near Aandalsnes. During the battle anti-aircraft gunners from the destroyer HMS *Auckland* shot down Ju 88 A-1, 4D+BT of 9./KG 30 killing its entire crew (Uffz. Willi Holz, Fw. Johann Ramming, Fw. Karl Eilert and Gefr. Kurt Voges). KG 30 had more luck on the following day when crews from II./KG 30 sank two British trawlers in the same area (HMS *Penn* and HMS *Hercules II*. Another op on that day was mounted by Ju 88s from III./LG 1, whose crews attacked railroad lines between Dombas and Verma and dropped bombs on Aalesund train station.

On Monday, April 22, a formation of 19 Ju 88s from KG 30 and III./LG 1 hit enemy shipping near Romdalsfjord sinking a small Norwegian

A formation of six Ju 88 A-1s flying over French countryside.

Ju 88 A-1 from 1./KG 30 photographed after a crash landing. Note the unit's insignia on the engine cowl.

vessel *Sigurd Jarl* and damaging the destroyer HMS *Jackal*, as well as the trawler HMS *Pelican*. One of the attacking Junkers bombers of 9./KG 30 was hit by AA fire from HMS *Pelican* and went down in flames killing the entire crew commanded by Fw. Hans Ramming.

Another anti-shipping operation took place on April 23 when a force of 25 Ju 88s appeared over waters just off Aandalsnes. Ju 88 L1+FS of 8./LG 1 received a direct hit and crashed on the north shore of Romdalsfjord. Uffz. Hans Mahnke was killed in the crash, while the remaining

Replacing Ju 88 engines under field conditions was a challenging operation for the ground crews.

An SC 250 hanging under the wing of a Ju 88 A-1 from KG 51.

crew members (Uffz. Ernst Röder, Uffz. Otto Horns, Uffz. Hermann Schmidt) were taken prisoner. Another Ju 88 (belonging to III./LG 1) was destroyed when it crashed minutes after take-off from its home plate.

At noon on April 25 a formation of 12 He 111s from II./LG 1 and six Ju 88s from III./LG 1 bombed the frozen lake Lesjaskog, which was used as a divert landing field by British fighters. Less than an hour later Ju 88s from 2./KG 30 escorted by a swarm of Bf 110s of I./ZG 76 and a swarm of Ju 88 Cs from (Z)./KG 30 attacked targets in the Dombas/Aandalsnes area. On their way back to base the crews of Bf 110s and Ju 88 Cs strafed the British aircraft parked at the frozen lake Lesjaskog destroying four Gloster Gladiator fighters. Destruction of one of the Gladiators was credited to Ogefr. Richard from (Z)./KG 30. In the late afternoon Ju 88 (L1+BR) from 7./LG 1 was lost while attacking enemy shipping near Stadlandet. Ofw. Friedrich Katzmaier's crew (Gefr. Gerhard Betzin, Fw. Heinrich Cordes and Gefr. Willi Holthausen) were captured by the Norwegian and then handed over to the British, who shipped them to a POW camp in England. The bomber fell victim to Blenheims, whose fighter variant was operated by No. 254 Sqn. The German raid inflicted heavy damage on the trawler HMS *Bradman*, whose crew was forced to ran the ship ashore. The Norwegian torpedo boat *Trygg*[19] was sank in the attack, while two other British trawlers (HMS *Bradman* and HMS *Hammond* sustained heavy damage.

At the same time a flight of three Ju 88s from 2./KG 30 launched an unsuccessful attack on the battleship HMS *Warspite* some 160 km off the coast of the Shetlands. The final Luftwaffe operation of the day took place in the evening when 56 He 111s from KG 26 and KG 54 accompanied by a force of 18 Ju 88s from KG 30 dropped bombs on the frozen lake Lesjaskog.

Two days later, on April 27 a formation of 34 Junkers Ju 88s mounted another anti-shipping operation west of Moldefjord. The Luftwaffe bombs sank a large Norwegian transport *Nyhaug* (4,044 tons) and left two other ships damaged (the tanker *Delius* – 6,065 tons and escort ship HMS *Black Swan*). Another operation launched that day was an attack by 19 He 111s from II./KG 26 and Kgr. 100 and 26 Ju 88s from KG 30 against targets in Aandalnsnes. During the raid Ju 88 crews came under attack by a flight of five Blackburn Skua fighters from No. 800 and 801 Sqn., but managed to escape unscathed.

In the morning of the following day, April 28 11 Ju 87 Rs from St.G 1, 12 He 111s from KG 26 and LG 1 and 16 Ju 88s from KG 30 attacked a flotilla of enemy warships and transports in

Ground personnel digging air raid trenches. Visible in the background is a Junkers Ju 88 A-1 from LG 1.

Aalesund and Aandalsnes fjords. The crews of KG 30 sank a small Norwegian coastal steamer *Brand IV*. During a noon sortie later that day the KG 30 bombers were jumped by Skua fighters of No. 800 Sqn. Following the fight the Ju 88 gunners reported a single air-to-air kill

During a strike on Namsos launched on April 29, 1940 a formation of 24 Ju 88s from LG 1 sank a Norwegian transport *Orland* (1,899 tons). In the early morning of April 30 a Ju 88 C from (Z)./KG 30 crashed while attempting to land at Stavanger after a sortie over Aandalsnes. Luckily, all crew members walked away from the crash without a scratch. In the afternoon five Ju 88 As from 2./KG 30, escorted by three Ju 88 Cs from (Z)./KG 30, attacked a Norwegian air base at Setnesmoen. Three Ju 88 C pilots (Lt. Pack, Ofw. Hermann and Uffz. Wiesmann) claimed the destruction of three Fokker C-V reconnaissance aircraft parked on the ground. On May 1, 1940 18 Ju 88s from II./KG 30 launched a series of unsuccessful attacks on British fleet withdrawing from Namsos. Later that day KG 30 received orders to return to Germany where the wing would operate as part of Luftflotte 2. Only 6./KG 30 and (Z)./KG 30 were to continue their operations over Norway.

On May 10, 1940 Ju 88 C-2, W.Nr. 0880133, 4D+FH made a forced landing near Storfjellet following an engine loss. The bomber's crew (Uffz. Albrecht Weimann, Uffz. Ernst Schulz and Gefr. Helmut Schwanz) were taken prisoner. The aircrews of (Z)./KG 30 flew a successful operation on May 16, 1940. While escorting a formation of eight He 111s of Kgr. 100 Oblt. Bönsch shot down a Skua fighter from No. 803 Sqn flown by Lt. Harris. Half an hour later the British fighters attacked eight Ju 88 A bombers from 6./KG 30 over Ototfjord and downed two of them. Junkers Ju 88 A-1, 4D+AP fell victim to the crew commanded by Mdspmn. R.W. Kearsley. The German bomber ditched in the fjord and all crew members managed to make it to the shore safely. However, the British fighter swooped in and strafed the German crew killing Fw. Ernst Kübler. Hptm. Günther Noll (Staffelkapitän) and Fw. Hubert Trimborn were captured and ended up in a POW camp in Canada. Fw. Willi Shell was also captured, but he returned to his unit in August 1940. The other Ju 88 A-1 (4D+KP) was shot down by Sub./Lt. R.A. Eaton flying Skua L3010 "8B" of No. 803 Sqn. The German bomber crash-landed south of Ototfjord, but the entire crew survived. The pilot, Lt. Heinrich

Ground personnel performing maintenance on a Junkers Jumo 211 B-1 engine.

Junkers Ju 88 from KG 51 undergoing maintenance.

Diermayer, was captured, while the remaining crew members (Gefr. Georg Schultz, Uffz. Anton Schloßer and Gefr. Heinrich Koch) managed to evade the enemy and returned to their unit on June 13, 1940.

On May 18, 1940 eight He 111s from II./KG 26 and three Ju 88s from 6./KG 30 attacked a British naval force in the North Sea. One of the PC 1000 bombs dropped by the Junkers crews hit a Royal Navy warship, penetrated her three decks, but failed to explode. The bomb killed two of the ship's crew and wounded another 26. The damaged ship immediately returned to Scapa Flow.

On May 20, 1940 Ju 88 Cs from (Z)./KG 30 flew en escort mission over Narvik in support of Ju 52/m3s. One of the aircraft (Ju 88 C-2, W.Nr. 0137, 4D+GH) suffered an engine failure over Vik forcing the crew to make an emergency landing. The bomber's pilot Fw. Eugen Kretschmer and radio operator Gefr. Konstantin Erhardt were taken prisoner, while the gunner Ogefr. Fritz Vogt managed to evade capture and rejoined his unit on June 10, 1940.

On Saturday, May 25 a Ju 88 from 6./KG 30 made a successful attack on the British troop ship HMS *Mashobra* near Harstad. The crew was forced to beach their damaged vessel to keep her from sinking. On May 28 F/O Lydall, a Hurricane pilot with No. 46 Sqn., shot down over Narvik a Ju 88 A from 6./KG 30 flown by Fw. Ernst Alf. One of the bomber's crew was killed in the attack, while the other three were captured. On June 7, 1940 Ju 88 C-1, W.Nr. 0142, 4D+MH, Stammkenzeichen VB+WC was shot down while attacking ground targets in the Narvik area. The bomber made a forced landing near Bjornefjell during which one crew member suffered minor injuries. On the following day the Allies completed evacuation of their troops from Narvik, which was a symbolic ending to the Norwegian campaign.

Western campaign - May 10 to June 22, 1940

While the fighting in Norway was still going on, the Germans launched their offensive on the western front in the early morning of May 10, 1940. In the first stage of the invasion, code-named *Fall Gelb* (Case Yellow), Army Group A was to advance through the Ardennes towards the English Channel, while in the north Army Group B was tasked with launching a frontal attack across Holland and Belgium. Army Group C was supposed to secure the southern flank of the German front.

The Luftwaffe's role included two main tasks. The first and the most important one was to neutralize enemy's air and to achieve air superiority over the battlefield, which would pave the way for the attacks on Allied transportation hubs and supply lines and allow the Luftwaffe to provide close air support to ground troops.

The Luftwaffe forces taking part in the French Campaign included the following Ju 88 units (unit name, CO, airfield, total number of aircraft/combat-ready aircraft):

Luftflotte 2, Gen.d.Fl. Albert Kesselring, Münster:
1.(F)/121 Hptm. Fischer Münster/Handorf, 3/2
2.(F)/122 Oblt. Schwartz Münster/Loddenheide, 5/4
3.(F)/122 Maj. Keienburg Münster/Loddenheide, 2/2
4.(F)122 Hptm. Pannwitz Goslar, 5/2

IV. Fliegerkorps Gen.d.Fl. Alfred Keller, Düsseldorf:
Stab (K)/LG 1, Obst. Alfred Bulowius, Düsseldorf, 1/0
II.(K)/LG 1, Maj. Kurt Dobratz, Düsseldorf, 32/4
Stab KG 30, Obstlt. Walter Loebel, Oldenburg, 2/2
I./KG 30, Hptm. Fritz Doench, Oldenburg, 34/25
II./KG 30, Hptm. Claus Hinkelbein, Oldenburg, 38/25
III./KG 30, Hptm. Arved Crüger, Marx, 30/20

Fliegerführer of.b.V. 2,Gen.Maj. Richard Putzier, Bremen:
III./KG 4, Maj. Neudörffer, Delmenhorst, 37/21

Luftflotte 3, Gen.d.Fl. Hugo Sperrle, Bad Orb:
3.(F)/121 Oblt. Knapp Frankfurt/Main, 3/2
4.(F)/121 Hptm. Kerber Gablingen, 2/1
1.(F)/123 Oblt. Sembritzky Langendiebach, 6/3
3.(F)/123 Hptm. Liebe-Piderit Gelnhausen, 3/2

V. Fliegerkorps, Gen.Lt. Robert Ritter von Greim, Gersthofen bei Augsburg:
Stab KG 51, Obst. Josef Kammhuber, Landsberg/Lech, 1/0
I./KG 51, Maj. Hans-Bruno Schulz-Heyn, Lechfeld, 23/7
II./KG 51, Maj. Winkler, München/Riem, 39/29

In total, the force included 237 Junkers Ju 88 A bombers, although only 133 of those were fully combat-ready. There were also 29 reconnaissance Ju 88 of which 18 were ready for action.

At that time the Ju 88 was still a new design and maintaining the type's high availability was not an easy task (as evidenced by the above table). Things were just as difficult with availability of aircrews. Most of the pilots had only a few days worth of experience on the type. One of them was Uffz. Peter Wilhelm Stahl, the future KG 30 ace: "I finally completed conversion to the Ju 88 today (April 30, 1940 – author's note) with the help of a factory test pilot named Barnickel. I was in the pilot's seat from the first flight. Flying the ship was considerably different from all the other types I had flown so far. At low speed the aircraft's flight controls weren't very effective, which made take-offs and landings very demanding. The ship was in fact a bit of a diva. She seemed to be aware of just how beautiful and interesting she was and she behaved accordingly. Without a slightest warning, she was fully capable of doing things that you'd least expect. She was especially capricious on take-off roll. You could keep her rolling straight as an arrow using the rudder until the moment the tail wheel rotated and the fuselage assumed horizontal attitude. That was something the diva didn't seem to like much. Without a hint of a warning, almost instantaneously, she would yaw to the left – a move that was hard to arrest if your control inputs were just a fraction of a second too late. On landing you had to maintain 250 km/h until the flare, which had to be done at 215 km/h – not much of a margin for error if you wanted to land the bird safely. However, after three circuits I felt completely at home in the cockpit and I was cleared to fly solo with only a flight engineer for company. This is where hard work and time spent in ground school really paid off. Unfortunately, most of the pilots still take it for granted and try to just coast through ground training hoping that things will somehow work out in the end..."[20]

The first Ju 88 combat sortie of the western campaign was flown by the aircrews of 7./KG 4, who laid mines on approaches to Dutch harbors on the night of May 9/10, 1940. Before dawn on May 10 Ju 88s from KG 30 went into action as well. Lt. Werner Baumbach remembers: "On May 9 I learned from a radio dispatch that I had been awarded the Knight's Cross. By the next morning we were in combat again. The night's events made it all clear. From that moment on, everything we did had just one purpose: to gain control over the key areas in Belgium and Holland and to maintain it until our ground forces came marching in through the borders. The only way to achieve such ambitious goals was to stick religiously to the time table of the operation.

Our task was to neutralize flak batteries deployed around airfields around Rotterdam,

Ju 88 A-1 of KG 51, which crashed in France in the spring of 1940.

the Hague and Delf. The dark, night skies was a great cover for our preparations. After a period of time we once again had small caliber bombs hung on the wings. Our ground crews worked tirelessly, which gave us, the aircrews, a few hours worth of sleep. Take-off was set for 04.30.

And then, once again, we were in the middle of a familiar whirlwind that resembled countless ops against the British. Squadron commanders passed operational updates to the flight leaders and several minutes later crews were driven to their machines in a duty crew car.

Before long the air was filled with the whine and thunder of engines operating at maximum power. The ground crews were quickly completing the final checks before reporting the readiness of their assigned machines to aircraft commanders. Then the airfield lighting was switched on. First to roll out of the ramp was the squadron commander's aircraft, which was followed by individual flights forming up behind him. The narrow taxiway was lit up by floodlights and trembled from the roar of our engines. The presence of a huge force, aching to be released, could be felt in the air.

When a predetermined signal was given, the squadron's CO began his take-off roll. Almost instantly, the other heavily-laden bombers also began to roll down the field. We all headed straight into the black, night sky. I took a quick glance at the clock – it was 04.31. Within minutes we were flying through the twilight sky. The day was May 10. In front of me was the flight led by Oblt. Beermann. We were all flying in a neat, tight formation using flames coming out of the exhaust stacks as points of reference. The formation was slowly gaining altitude to reach our predetermined attack height. It was getting lighter and lighter every minute. Down below we could see the lowlands and marshes of the Emsland region covered with a blanket of ground fog.

We were over Holland. Menz put away his map and took a fix of our position using the ADF. 'We're on course', he concluded.

I turned on the autopilot, which gave us all an opportunity to share a tin of biscuits. Noll's flight to our left was slowly catching up with us and we were cruising at the same altitude. We overflew several small Dutch towns and could see flashes of anti-aircraft guns popping out here and there. However, the puffs of the exploding shells were all way below us, so we didn't pay much attention to them. In the meantime, while the ground fog was slowly burning out, we flew into a thick cloud and almost lost sight of the Dutch countryside below. Inside the cockpit we could hear the steady hum of the engines. The hum, the rhythm that we heard was the sound of war, the sound of modern air warfare. For a brief moment we were all quiet, immersed in our own thoughts.

Before long it was back to reality. We had just arrived over the target area. Straight ahead we could see puffs of exploding flak shells. It looked like heavy artillery was ready for our arrival. We still couldn't see our target as it remained obscured by heavy fog. But that was our cover, too as at that moment the gunner yelled: 'Twin-fuselage enemy fighter, below and to the left!'

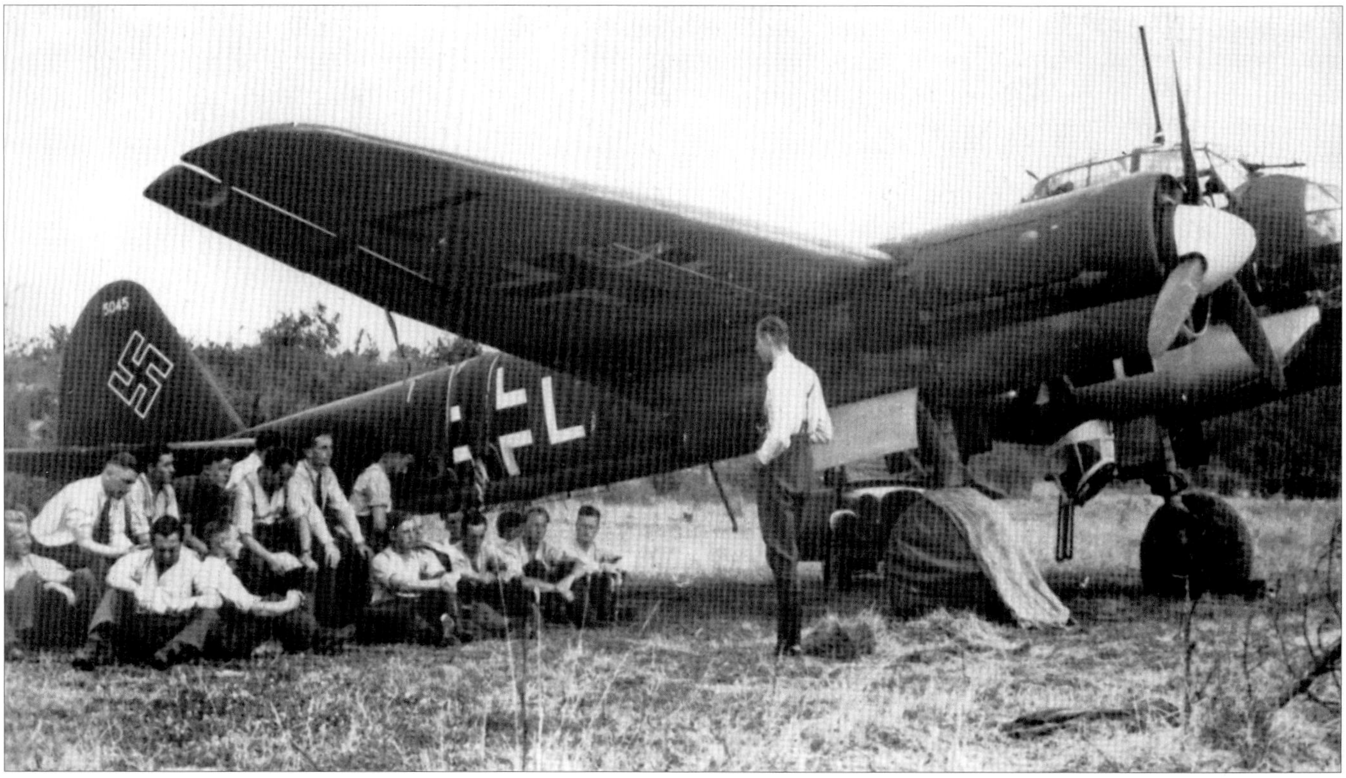

Junkers Ju 88 A-1, W.Nr. 5045, 9K+LR from 7./KG 51 parked at one of the airfields in France. Of note are white propeller spinners.

Aircrews in mission briefing. Note unusual paint scheme of the propeller spinners.

I knew those Dutch machines. I made a wide turn to the right to take cover in the bank of thick fog. We have to stay well clear of such unwanted surprises, I thought. Four sets of eyes were scanning the milky fog while the navigator, radio operator and gunner clutched their machine guns. I took a quick look at their faces: they were all great guys.

We took stock of the situation and got our bearings straight. We should be arriving over the target any minute now. All around us the sky was filled with flak. This was a sight we all remembered very well from our previous attacks on British ships.

When the radio operator screamed 'Fighters!', I briefly lost sight of the target. Just seconds later another, much calmer call came in: 'Friendlies, Me 109s, Me 109s!' Right on schedule, our own fighters and destroyer aircraft took us under their wings over the target area and swept up the airspace in front of us. The twin-fuselage fighter, whose pilots liked to think they were 'Messerschmitt eaters', was nowhere to be seen.

Menz had already set up the bombsight and armed the bomb ejectors. 'All's ready!', he reported. We were all set to attack. I passed the attack signal to my wingmen and we commenced the strike. Our formation flight terminated directly over the target as it was now up to each aircraft commander to assume the most favorable attack position. Our motto was: 'Approach together, strike alone.'

As the aircraft in front of me were going into their dives, I took a good look at our small target on the edge of the field – a flak emplacement in a small farmyard. Located on a small hill, the gun had a very good firing position. We had to neutralize it before it could ruin the day for the paratroopers coming in behind us. I saw the first bombs explode along the southern edge of the field.

I decided to make a low-level pass, making a wide turn and then coming in from the west. I arrived over the target in a descending glide before dropping the nose and going into a dive. Within seconds the nose was pointing directly at the target and the farmyard was getting bigger and bigger in the bombsight. I thumbed the pickle button on the control column. 'Bombs away!', announced the gunner lying prone in the ventral tub. Seconds later came his next report, delivered with total calmness and composure: 'Bombs on target!'

Egressing the target area I saw the farm buildings fold like a house of cards. Anti-aircraft fire ceased and we were on our way back to base. No sooner did I say: 'Watch out for enemy fighters', than we came under attack by the twin-fuselage fighter. Thies, my radio operator, spotted the bandit in the nick of time and we opened up on him with all we had. I was desperately trying to find some clouds, but the sky was clear. The only way out was to bring the nose up sharply and then dive for the deck. 'Goddamn,

The crew of a Ju 88 A-1 from KG 51 heading out to their aircraft before another combat mission.

the piece of shit is jammed!', I suddenly heard. Willem Braun, who had just reached for a fresh magazine stowed behind my seat, immediately pointed the barrel of his machine gun upwards. I took a quick look over my shoulder and saw the enemy fighter right next to us. I could clearly see the pilot and the machine's national markings. Both our machine guns rattled again, Thies firing like mad. We hit the deck and managed to shake the guy off. He didn't try to chase us. I'm sure one of the 109s got him.

At the same time we saw our paratroopers, who began their assault right on schedule.

On our way back we watched the grey columns of ground troops rolling forward across the Dutch countryside. The victorious western campaign was underway."[21]

One of the Ju 88s from 8./LG 1, flown by Fw. Ernst Schade, was downed over Belgium by F/O Ken Blair of the RAF's No. 85 Sqn. This is how the British pilot remembered the encounter: "The enemy aircraft opened fire from a range of 1,000 yards. I closed in to within 400 yards and squeezed off three four-second bursts. I then saw thick smoke coming from the starboard engine before the machine nosed over and disappeared in a layer of ground fog."[22]

KG 30 lost as many as eleven Ju 88 A-1s. 4D+CB from Stab KG 30 came under attack by a Fokker D XXI of 1-II-1 and crash-landed near Rozendaalse killing Uffz. F. Lassen. The remaining crew members (Fw. G. Hartmann, Fw. K. Soldau and Gefr. E. Gieseler) were captured; Ju 88 A-2, W.Nr. 2017, 4D+IT from 1./KG 30 was hit by AA fire south of Rotterdam and crashed when the crew attempted a forced landing. An exploding AAA shell killed Uffz. S. Weiß, while Oblt. R. von Wartenberg and Uffz. Müller were both killed in the crash. Gefr. D. Falcke suffered severe burns and was taken prisoner; Ju 88 A-2, 4D+FT of 1./KG 30 made an emergency landing near Zevenhuizen with battle damage to one of the engines. The entire crew (Oblt. Wülknitz, Ogefr. H.-W. Abt, Ogefr. H. Borchers and Gefr. G. Manthei) ended up in a POW camp; Ju 88 A from 3./KG 30 was severely damaged by AAA fire during a low-level attack on Soesterberg and was subsequently destroyed while attempting to land at Rheine. The crew walked away from the crash; another Ju 88 A from 3./KG 30 sustained severe damage following an attack by a Fokker G 1 of 3-II-1 just east of Rotterdam. The bomber was lost while trying to land at Rhein, killing Uffz. H. Schreiber and Gefr. E. Schröder. Uffz. H. Hobik and Ogefr. F. Baumgartner suffered only minor injuries; Ju 88 A-2 from 3./KG 30 was shot down by AAA fire during a diving attack on Waalhaven and crashed near Ouderijn killing , Oblt. H.-W. Magnussen, Uffz. K. Roscher and Uffz. G. Richter. Ofw. L. Erdmüller, who suffered injuries in the crash, was captured and became a POW; Ju 88 A from 4./KG 30 did make it back to base in spite of severe battle damage sustained over Waalhaven, but it was subsequently written off. Two of the crew members, Ogefr. O. Zahnke and Ofw. H. Bloeck, suffered wounds; Ju 88 A, 4D+AM from 4./KG 30 was shot down by a Dutch fighter of 1-V-2 while attacking AAA emplacements near the Hague. Ogefr. P. Müller was killed in action, while the remaining crew members were captured (critically wounded Ofw. A. Stahl, Staffelkapitän Oblt. T. Beermann and Ofw. E. Maar); Ju 88 A z 5./KG 30 was lost on take-off from Oldenburg. The bomber hit a flak emplacement killing two of the gunners and wounding another eight. One of the crew members (Gefr. A. Sattelmacher) suffered serious injuries in the crash; Ju 88 A-2, 4D+KR z 7./KG 30 received a direct hit from a flak shell and exploded in mid-air over Nijmeegen killing the entire crew (Lt. H. Begemann, Fw. A. Herb, Uffz. W. Haiges and Flg. H. Müller); another Ju 88 A-1 crashed on take-off from Marx. Uffz. H. Landsberg, Ogefr. H. Hoetz and Gefr. R. Alksaat died in the crash, while Uffz. H. Haase suffered serious injuries.

III./KG 4 also paid a heavy price losing five of its bombers: Ju 88 A-1, W.Nr. 3040, 5J+HR from 7./KG 4 crashed minutes after take-off from Delmenhorst killing the crew (Lt. H.-J. Schwartz, Ofw. W. Voigt, Ogefr. H. Wilshaus and Uffz. W. Großmann); Ju 88 A-1 from 8./KG 4 was shot down by friendly AAA fire over Arlon. Fw. F. Wehrkoff, Fw. W. Müller and Uffz. Kunath were all killed in action; Ju 88 A-1, W.Nr. 3041, 5J+GT from 9./KG 4 was lost over Schiphol following an attack by a Fokker D XXI. Gefr. F. Strada was killed, while Oblt. F.-K. Rinck, Ofw. W. Rüther and Gefr. F. Wissing were captured; another bomber shot down over Schiphol was Ju 88 A-1, W.Nr. 6018, 5J+AT of 9./KG 4. The entire crew (Lt. G. Hinsch, Ogefr. A. Fölsch, Ogefr. K. Danzmair and Uffz. Herzog) were killed in action.

Ju 88s also took part in heavy fighting over France, which started within the first few days of the German offensive. One of the pilots involved in those battles was Fw. Friedrich Aufdemkamp from II./KG 51: "On May 9, 1940 we deployed to München-Riem airfield. The aircraft, fuelled and ready for action, were dispersed along the field's perimeter. All the machine guns were checked and doubled-checked and loaded with live ammunition. Finally, the bombs were hung under the wings. The typical load-out was four 250 kg bombs and twenty 50 kg weapons.

In the meantime the aircrews were briefed on the target area and attack tactics. As far as I remember, it was Dijon airfield. Take-off was scheduled for 04.00 on the following day. We were supposed to cross the French – German border at first light. (...)

And then, finally, we were woken up at an ungodly hour on May 10, 1940. After we topped up breakfast with additional food rations (half a liter of milk or porridge, an egg, black coffee and a Scho-Ka-Cola bar), we went in for the final brief. The aircraft were to take-off at short intervals, rendezvous and form up, and then head out towards France. The crews were given a bit of time to calculate their outbound course and estimated time over target. When the scheduled take-off time approached, we moved out to our aircraft. Hptm. Berlin, the squadron's CO, taxied out first. Following him, in a tight column, were aircraft from 6, 5 and 4 Staffel. Since I was assigned a position in the CO's flight, I followed right behind his machine. Once I lined up on the runway I took a look through the side window and saw a scene that made my heart race, but was at the same time awe-inspiring. The long line of dark silhouettes behind me was dotted with red, green and white flickers of position lights, while the ground was flooded with streams of bright light produced by taxi lights in the aircraft's wings and fuselages. The exhaust stacks of the mighty Jumo 211 engines were shooting pale blue flames. That night parade of thirty machines, each loaded with 1,000 kg of fragmentation bombs, looked like an illuminated beast bringing death and destruction. But there was no time for daydreaming. Hptm Berlin's Ju 88 was ready for departure and soon I received a signal to move closer as well. It all suddenly became dead serious.

Beyond the illuminated departure end of the runway I could see nothing, except a dark line of a forest. Foggy conditions and totally blacked-out city made it impossible to make out the horizon. I was about to perform an instrument take-off.

The CO had just been given a green light. My 9K+JP was also ready, so I just sat there watching the boss's roll. He smoothly rotated the aircraft and began a gentle climb. When he was at about 30 – 40 m over the field, I realized, to my horror, that the aircraft was losing height. Within seconds the bomber dropped the right wing, which hit the ground trailing a streak of

This Ju 88 A-1 from KG 51 carrying four SC 250 bombs is ready for departure.

Ju 88 A-1 of KG 51 in flight over France.

fire behind it. After a while the machine came to rest and turned into a huge, burning torch.

'That's not a good beginning', noted Hermann Schäfer breaking the silence in which we watched the burning wreckage. 'Did anybody out there have any chance to survive that?' There was an inevitable delay following the mishap, but finally our crew got the green light. I pushed the throttles and heard the satisfying roar of both engines. Despite the heavy bomb load and 4,000 liters of fuel in the tanks, the aircraft was accelerating nicely. After around 80 – 100 meters of ground roll, the left main landing gear tire blew and the machine immediately lurched to the left, departing the runway. I managed to continue rolling to stay clear of the aircraft behind me. All crews were soon airborne, but for us that first mission was already over.

Hptm. Berlin and two of his crew suffered severe burns, but survived the crash. The radio operator could not be saved. For the KG 51 crews that first combat mission was not particularly successful."[23]

Ju 88 A-1 from 8./LG 1 was shot down over Hermaville, north-west of Arras. The gunner, Gefr. Heinrich Maiborn, bailed out safely and was subsequently captured, but the other crew members (Fw. E. Schade, Uffz. G. Kronewitz and Uffz. L. Hechmann) were all listed as killed in action.

On the following day, May 11, 1940 Junkers Ju 88s provided close air support to German ground forces fighting in Belgium and Holland. A number of reconnaissance sorties were flown as well. In the early afternoon a flight of Spitfires from No. 19 Sqn. patrolling just off the coast of Holland shot down a Junkers Ju 88 A-1 belonging to 3.(F)/122. The bomber crashed into the North Sea near Terschelling, killing Uffz. H. Kunze. The remaining crew members (Lt. F. Meyer, Uffz. R. Krulich and Uffz. W. Höbel) survived and became POWs.

The base of II./KG 30 at Oldenburg was also a busy place that day as remembered by Uffz. Peter Wolfgang Stahl: "After a night out in Berlin we arrived back at the base at around noon. The place was buzzing with activity since the operations against Holland had just started the night before. Ju 88s with eagle insignia on their noses were taking off or landing almost non-stop. The command post saw an endless procession of aircrews – coming and going. Wherever you looked, there were oil-stained flight suits, yellow life vests, fur-lined boots and yellow flight helmet covers. (...)

An NCO stops to say, with a broad smile, that he returned to base with 70 bullet holes all over his aircraft, which didn't keep him from completing the mission. One of the machines taxis in right in front of the command post, which is where I'm standing, looking like a Red Cross representative or a member of a propaganda detachment. The hatch flies open and I can see a bloodied hand just hanging there lifelessly. Within seconds medics pull the seriously wounded gunner from his station in the ventral tub. He immediately drops to the ground before being whisked away in an ambulance. The poor guy had been shot in the neck. Later the body of the observer is pulled out from the bullet-riddled machine."[24]

Among the aircrews flying combat ops that day was Fw. Friedrich Aufdemkamp from II./KG 51: "On May 11, 1940 our squadron was to bomb a French airfield at Malbouhans near Lure, on the west side of the Vosges. We took off in a flight formation, which was not a problem considering the size of the airfield at München-Riem. We were understandably stressed out and apprehensive as we crammed into the cockpit. After all, that was our first combat mission. After we had formed up we began a climb to 4,000 meters. It was very disappointing to all of

Maintenance of one of the Junkers Jumo 211 B engines mounted on the Ju 88 A-1.

us that the fully combat-loaded Ju 88 could only manage about 250 km/h while flying in a formation.

After about an hour we were over the Schwarzwald. So far there were no surprises, except one abort probably caused by an engine failure. My crew were busy getting their guns ready. I told them to keep an eye out for enemy fighters. The tension reached its climax when we crossed the Rhine just south of Colmar and entered French airspace. Lying prone in the ventral tub, Günther Vossen thought he saw flak shells exploding far behind us. But the air around us was perfectly quiet. Suddenly all hell broke loose. As far as I could see there were puffs of exploding AAA shells, some frighteningly close to us.

The formation, nice and tight until we reached the French border, now scattered all over the sky like a flock of scared chicken. Well, they did say: maintain loose formation when engaged by AAA, but keep tight when attacked by fighters. In the meantime the flak fire stopped almost as abruptly as it had begun. Now only a shell or two reached our cursing altitude and exploded harmlessly in the distance. All the same, we couldn't let our guard down since the French fighters could show up any minute now. We tightened up the formation and waited, but that day the pleasure of looking the French fighter pilots in the eye was not to be. In the meantime we were already some 100 km into the French territory, west of the Vosges. Hermann, who had been keeping a keen eye on our progress, told us we were approaching the target area. he had already set up the bombsight and armed the bombs. The CO's flight was far ahead of us and turned north-west over the town of Lure. After a few more minutes we saw our target. Here we go, I thought. Soon I saw the first bombers peeling off the formation one by one and diving in on the target. We waited for the CO's signal to commence our bomb run.

A Ju 88 pilot had a number of checks to perform before commencing a dive: set the required bomb release altitude on the dive altimeter, close the radiator flaps, set the prop pitch and set the elevator trim to "dive" position. Once the aircraft approached its stall speed and when dive brakes deployed, you pulled back on the throttles and went into a dive.

I initially selected the middle of the field as my aim point. Before I transitioned to the bomb sight I saw the first bombs exploding down below. I than caught a few aircraft parked on the field in my bombsight and put the aircraft into a 70° dive. We were swooping down on the target at 550 km/h. When I heard the altitude warning tone I pressed the red bomb release button on the control column. At the same time the automatic dive recovery system kicked in pulling the aircraft up in a steep climb. Under tremendous g's my body was squeezed into the seat and I momentarily blacked out. Now was the time to quickly stow the dive brakes, add power, re-trim the aircraft and adjust prop pitch.

Paul Kolodzic and Günther Vossen who were in the best position to observe the impact of our bombs reported that they were all on target. They may have been right, since the entire field was obscured by clouds of smoke. However, there was only one thing that I cared about at that point: get out of there as soon as possible and rejoin the formation. I knew that we were nothing but sitting ducks for enemy fighters. Fortunately, they were nowhere to be seen. The AAA did open up on us, but the gunners soon ceased fire realizing they couldn't get us,

Even though our ship was now 1,000 kg lighter, it took us a while to rejoin the formation. There were still a few stragglers behind us, some flying solo, some in two-ship elements, all trying to catch up with us. When we reached the Rhine lowlands AAA fire started again. The formation, by that time cruising at 4,000 meters, once more loosened up. Once we reached the Rhine, the flak batteries went quiet.

Until that moment there had been very little radio chatter, since all crews were busy either looking out for enemy fighters or dodging the AAA fire. Now that the worst was over, there was a lot of happy banter going on. We were all happy to have made it in one piece through our baptism of fire and couldn't wait to share the joy with others.

We arrived overhead our base at Fürstenfeldbruck in a very loose formation and landed at around 14,00. It was then time for all the crews to go into the debrief where any claims could also be reported. We soon found out that all aircraft returned safely to base."[25]

Tasking for the Luftwaffe remained unchanged on May 12, 1940. The key objectives were elimination of enemy air and close air support for ground forces fighting in Holland and Belgium. On that day six Defiant fighters from No. 264 Sqn. along with six Spitfires from No. 66 Sqn. intercepted a Ju 88 A from 7./LG 1 off the Dutch coast. The German crew were attacking an Allied destroyer on approach to Ijmuiden. Despite being greatly outnumbered, the Germans managed to shake off the fighters escaping with only minor damage to their aircraft and one wounded crew member. Less lucky was the crew of Ju 88, 7A+CH from 1.(F)/121 who were shot down over Tielt. The pilot, Fw. Kurt Denner, was killed in the attack, while the other crew members (Lt. E. Bruns, Gefr. G. Damazyn and Uffz. A. Steinbock) all ended up in POW camps.

Fw. Friedrich Aufdemkamp found himself in the heat of battle once again: "On May 12, 1940 II./KG 51 launched for an attack on the Verun – Sedan railroad line. Our squadron (6. Staffel) was tasked with bombing a tunnel near the city of Stenay. In the early morning we moved from Fürstenfeldbruck back to München-Riem where each aircraft was loaded with four 250 kg bombs. We launched at 13.00. As soon as we crossed the French border we encountered AAA fire of varying intensity. Some of the shells, just like the day before, burst dangerously close to us. I could almost smell cordite in the cockpit. Luckily, none of our aircraft got hit. Over the target I botched it: having missed the stalling point I was forced to make another approach. This time I attacked the target from 800 meters dropping the bombs that exploded next to the

This Junkers Ju 88 A-1, 4D+AA, was the personal mount of Geschwaderkommodore KG 30, Obstlt. Walter Loebel. The tips of the propeller spinners are painted light green.

entrance to the tunnel. Unfortunately, we were unable to assess damage to the target.

After the attack I attempted to rejoin the rest of 6. Staffel, but couldn't get eyes on any of our aircraft. Having no other options, I asked Hermann for the direct course back to base. After two or three minutes I heard Günther scream: 'Two fighters, right behind us, closing in fast!' Paul also spotted the fighters, but none of us could tell for sure if they were friend or foe. I pushed the throttles forward to gain some speed. At that moment Hermann Schäfer yelled: 'They're French, probably Moranes! I can see the blue, white and red roundels! They're flying away now!'

'They're back!', screamed Hermann again, 'but it looks like they're keeping well clear of us!' Paul Kolodzic opened fire. Also Günther Vossen kept squeezing off short bursts from his machine gun. Both fighters made a series of evasive maneuvers and were really hard to hit. One of them got right behind us and followed our bomber some 100 m off to our left. I could clearly see the pilot sitting in the cockpit. 'Watch out, they're trying to catch us in the pincer!', warned Hermann Schäfer. I immediately firewalled the throttles and threw the aircraft straight into the guy flying on our left. The maneuver paid off handsomely since Paul was now in a perfect firing position and promptly opened up on the Morane at optimum range. Suddenly I heard him scream: 'Man, I got him! He's breaking off trailing smoke!' At that very instant the other fighter also broke off and headed out towards his wounded comrade. Before long they were both gone.

The joy and excited chatter that erupted after our lucky escape didn't last long, because we once again came under attack by heavy flak guns. The fire soon ceased and we arrived over Fürstenfeldbruck without any further incidents. Before we landed we buzzed the field rocking the wings to celebrate our air-to-air kill.

During the post-flight debrief we described the fighter attack and claimed one probable Morane kill. Later that day my crew and I were told to report to the command post, where general Ritter von Greim congratulated us and pinned the Iron Cross 2nd class on our flight jackets. I may have blushed a bit as I believed my achievements did not merit the award."[26]

On that day Junkers Ju 88 A-1 belonging to 5./KG 30 was shot down by Defiant fighters of No. 264 Sqn. and Spitfires from No. 66 Sqn. near Ijmuiden while attacking Allied ships. The bomber crashed at Hazerswoude at 14.30 killing Flg. S. Lahm and Uffz. W. Barowka. Fw. T. Schwarz died of his wounds in hospital on May 21, 1940, while Oblt. H. Schneider suffered only minor injuries and became a POW.

On May 13, 1940 German panzer units broke through the front at Sedan, which sealed the Allies fate in Belgium, despite counteroffensive attempts both on the ground and in the air. On the following day the Dutch began ceasefire negotiations. Under those circumstances II./KG 30 could be withdrawn from frontline operations to allow time for additional training that the unit's aircrews were to undergo at Perleberg. Before that happened, a Ju 88 A-2 from 8./KG 30 was lost to enemy AAA over Waddenzee. The bomber's crew (Lt. A. Kleinmacher, Uffz. M. Mamein, Gefr. W. Gröbel and Gefr. O. Beyer) were all captured by the Allies.

Junkers Ju 88 A "on top" during a combat sortie.

The only homogenous unit that still operated Ju 88s over the western front was now KG 51. On May 14, 1940 a Polish pilot ppor. Erwin Kawnik, flying Morane M.S. 406 from GC III/6, shot down a Junkers Ju 88 A-1, W.Nr. 4008, 9K+EL of 3./KG 51. The bomber's pilot, Fw. Friedrich Wurth, as well as the other crew members (Fw. R. Buck, Uffz. G. Pfeuffer and Uffz. H. Seidel) were captured. Another Ju 88 A from 3./KG 51 was lost during a raid on Dijon airfield. The aircraft was shot down by French fighters from GC III/6 and crashed near Zweifalten. Flg. K. Hinterhofer survived the attack despite suffering wounds, but the other crew members were all killed in action (Uffz. F. Kröll, Uffz. K. Schneider and Gefr. P. Holtfurth).

While bombing the train station at Revigny Junkers Ju 88 A-1, 9K+CS from 8./KG 51 was attacked by Dewoitine D.520 from GC I/3, which resulted in the bomber's crash landing near Neuhausen. The aircraft was completely destroyed, but only one crew member (Flg. F.-K. Mayr) suffered (minor) injuries.

Also lost on the same day were two reconnaissance Ju 88s. The first one (Ju 88 A-1, W.Nr. 0165, 7A+BH of 1.(F)/121) was shot down by a Hurricane from No. 504 Sqn. and crashed near Albert Canal killing the entire crew (Oblt.

Heinz Spillmann, Ogefr. R. Schnegotzki, Uffz. W. Colleseus and Uffz. W. Gers). The second bomber (Ju 88 A-1, F6+BL from 3.(F)/122) was heavily damaged by F/O Bill Fowler flying Hurricane (P2622) of No. 615 Sqn. The Junkers, its starboard engine shot up to pieces, made a forced landing near Aalst. Uffz. W. Reissmann was killed in action, while Uffz. Uffz. Erwin Maxrath, Fw.F. Küttner and Fw. E. Lauterbach were all taken prisoner.

On May 15, 1940 Ju 88 A-2 from 2./KG 30 was shot down while on a low level sortie against Allied shipping near Vlissingen. Flg. A. Trinkl survived and was picked up by a Dutch transport, while Uffz. L. Wengert, Uffz. M. Stampfer and Ogefr. J. Wunder were killed in action.

Oberst Josef Kammhuber, the CO of KG 51 and the man behind the Reich's night air defense system, remembers: "On May 16, 1940, six days after the western offensive had begun, I received a phone call from the Luftwaffe's Commander-in-Chief. His orders for my wing was to mount an immediate operation against railroad infrastructure in the Nancy-Epinal-Vesoul area. Göring insisted that the operation was an absolute priority and he expected my wing to perform to the highest standard, given that we operated the newest Ju 88 dive bombers. He had every right to expect nothing short of excellence from us, but only if the weather conditions were favorable. So what was the weather doing?

I got a phone call from our met people later that morning and here is what they had to say: '7/10 cloud cover over the Schwarzwald, tops at 5,000 meters. On the other side of the Schwarzwald cloud cover improving to 2/10 with good visibility.' That meant we would have to find the proverbial hole in the clouds over the Schwarzwald, since a long duration flight over that area without visual contact with the ground was too much of a risk. The time between Ju 88's introduction and its operational debut was so short that there was no time for proper instrument training for the aircrews. It was therefore obvious that there would have been inevitable losses if the entire wing were to penetrate the thick overcast in instrument conditions. Just two days earlier the left wing 'Kaczmarek'[27] in my wing's CO flight lost spatial orientation while punching through a relatively benign looking cloud layer and crashed killing the entire crew. I still have a vivid memory of an aircraft popping out of the clouds on my left in inverted attitude and disappearing in another cloud layer down below.

I took off at around noon leading the wing's Stab flight and Lechfeld-based I. Gruppe. We then rendezvoused with part of II. Gruppe who had already converted to the Ju 88. Once formed up, we headed west. Even from a distance I could tell that the weather had seriously deteriorated and there were absolutely no gaps in the dark, overcast skies over Schwarzwald. Nonetheless, I pressed on and decided to take a closer look.

In the meantime I climbed to 3,000 meters hoping to stay on top of the cloud layer. But that proved to be impossible as well. Looking up, it appeared that the clouds had no visible tops. In fact the overcast extended all the way up to 5,000 meters, which was the absolute limit for our Ju 88s, each carrying four 500 kg bombs. What should we do? Being aware of my crews in-

Ju 88 A, W.Nr. 7877, 9K+EL of 3./KG 51.

strument skills I knew that leading the wing into the clouds would result in huge losses even before we reached the target area. And then, if the weather hadn't improved, the flight back to base would have cost more lives and losses in equipment, catastrophic for the entire wing. Should I then abort? Should I turn around and return to base, even if the operation was ordered by the Luftwaffe's Commander-in-Chief?

It was a hard nut to crack for a Geschwader commander, especially one that was fully aware of the Reichsmarschall's volatile character! But I could still see the faces of those crews that stood in front of me that morning waiting for their orders; I thought about their loved ones and immeasurable grief they would all have to go through if those flyers hadn't returned home. That's when I made a decision to abort. Let the brass whine and bellyache all they want, I thought. My people were more dear to me and more important than any amount of anger at the top.

However, I didn't want to be accused of personal cowardice – which in those days was common practice – so I decided to continue the mission on my own. I ordered the rest of the wing to return to base and to jettison their unarmed munitions over a predetermined area at Lech-Auen. I then continued to fly through the thick overcast on instruments. As I got my instrument instructor rating back in the summer of 1939 I could trust my own skills in the Ju 88, although I had no actual instrument time on the type.

After about 30 minutes the overcast disappeared and I was flying over the Rhine lowlands at 5,000 meters. Lying ahead were the Vogeses and clear skies. If I only had the entire wing with me, I thought. I was even a bit angry with myself for issuing the abort order, but it had already been done and dusted.

We proceeded to the target area straight as an arrow. I then put the bomber into a dive, released all four 500 kg bombs and photographed the attack's results. We than loitered for a bit over the target area to better asses the damage. Having taken a few more pictures and completed the notes that might be helpful in the debrief, we turned east without encountering the slightest sign of resistance. I have to say we all felt a bit sheepish that the whole thing went so smoothly.

When we were once again over the lowlands the thick, black overcast was still sitting on top of the Schwarzwald. No matter, the mission objectives had been achieved, nothing could go wrong now. So we flew into the clouds setting course for our base at Lechfeld. In the clouds we encountered very heavy turbulence, which tossed our Ju 88 around like a toy. About ten minutes into that rollercoaster ride the autopilot system went out! I wasn't ready for that. It happened very abruptly and at the worst possible moment, when the machine hit extremely heavy turbulence. It made it impossible to smoothly transition from auto flight to hand flying.

Turn indicator needle all the way to the left, the ball on the right; artificial horizon showing a steep, climbing turn. Before I could react, the airspeed began to increase rapidly with the altimeter winding down like crazy. The airspeed

SC 1000 bombs were very dangerous weapons against the Royal Navy ships. Visible in the background is a Ju 88 A-1 of KG 51.

The nemesis of Ju 88 aircrews fighting over Dunkirk were the British Supermarine Spitfire fighters.

indicator went off the scale and stopped at 720 km/h. We're bailing out!

My radio operator, Lt. Lüderitz, murmured something to the effect: 'Nice flying, let's get the hell out of here!' The observer, , Lt.d.R. Krösch, seemed to be mesmerized by the altimeter, while our flight engineer, Uffz. Schneider, was struggling with the Bola hatch. He couldn't get it open as the high g's pressed us all into our seats. At that moment it dawned on me that probably the same thing happened to the poor 'Kaczmarek' a couple of days before. But at that very moment that was pretty useless knowledge. What good was it for us as we ourselves were hurtling down towards the ground like a rock?

Well, all we could do was to sit there and wait for death, which would inevitably come within the next few seconds. But it didn't. Low over the ground we popped out of the clouds and, using every last bit of strength, I managed to recover from the deadly dive.

To this day I can't figure out how I managed to do it. Next thing I remember was flying straight and level, low over a railroad line. There was a train below and we could see passengers waving to us enthusiastically. Well, we weren't all that enthusiastic ourselves. It was raining cats and dogs and mountains to the left and to the right were completely obscured by low clouds. We were lucky the whole thing happened over the lowest part of the valley near Donaueschingen. The cloud base was still very low all the way back to Lechfeld, which forced me to fly most of the leg on instruments.

The good thing was that I no longer felt angry with myself for ordering the rest of the wing to return to base. I don't know how many of the crews wouldn't have returned if I hadn't given that order.

There were two important lessons to learn from that experience, both critical for the entire wing.

The original auto flight system installed in the Ju 88 (Ascania-Zwei-Achs-Steuerung) wasn't robust enough for the type and, following our 'adventure', it was promptly improved eliminating future disconnect problems.

The second lesson was very reassuring. The robust and sturdy design of the Ju 88's wings passed the test of a brutal, instinctive dive recovery that I subjected them to. On other types the wings would have probably just folded and broken off in flight. All that the ground crews could find on my Ju 88 were small cracks in the wing's skin panels, nothing more. As a precaution, the wings were replaced and I flew the same aircraft again just a few days later."[28]

On that day three Ju 88s of I./KG 30 were shot down over Belgium. The first one was flown by the CO of 2./KG 30 Hptm. Karl Hielscher. The bomber was first damaged by AAA over Ninove and then finished off by fighters as the crew was trying to make an emergency landing near Wavre. Hptm. Hielscher suffered only minor wounds, but the rest of the crew perished in the crash (Ofw. H. Strohmeier, Uffz. W. Grazce and Uffz. S. Sempert). The second aircraft, Ju 88 A-2, 4D+CL belonging to 3./KG 30 came under attack by P/O Parsons from No. 504 Sqn. west of Brussels, who shot up the bomber's starboard engine. Minutes later the Junkers was hit by flak and crash landed near Ninove. Oblt. Rudolf Bürner and other crew members (Uffz. A. Henkel, Uffz. K. Schmitz and Gefr. D. Haase) survived and were captured by the Allies. The third machine lost that day was Ju 88 A from 1./KG 30, which was downed by AAA and crashed east of Ramsdonk, near the Dendermonde- Mechelen railroad line. Hptm. E. von Loesch, Fw. W. Steffen, Uffz Feyen and Gefr. K.-H. Teschner were all killed in the crash.

On Friday, May 17, 1940 F/L Jim Sanders from No. 615 Sqn. shot down Ju 88 A-1, L1+AR of 7./LG 1, which crashed near Fines-les-Raches. Critically wounded Staffelkapitän, Oblt. Ernst Schwarz was pulled from the wreckage and taken to hospital in Lille, but he died of his injuries on May 26, 1940. The remaining crew members

(Ofw. B. Hinke, Ofw. A. Dudek and Gefr. G. Salzmann) perished in the crash.

Another Ju 88 went down under heavy AAA fire on May 18 while attacking Allied ships at Dunkirk. The bomber (Ju 88 A from 8./KG 30) crashed killing Ogefr. T. König and Gefr. W. Mebus. The other two crew members (Lt. S.Jung and Flg. O. Barthels) suffered wounds and were captured by the Allies. III./LG 1 lost one of their Ju 88s at 16.45 over Amiens following an attack by F/O Connors of No. 111 Sqn. The crew was rescued by German ground troops.

On Sunday, May 19, 1940 Junkers Ju 88 A-2, W.Nr. 5035, 5J+FR from 7./KG 4 was on a reconnaissance mission when it came under heavy AAA fire. The damaged machine crash landed between Ostend and Boulogne. The crew walked away from the crash, although all suffered various injuries. The aircraft commander was Hptm. Erich Bloedorn, Staffelkapitän 7./KG 4, who was at the same time temporarily in charge of III./KG 4. The other crew members were Ofw. H. Saure, Fw. W. Stechardt and Uffz. R. Schaaf. Another aircraft lost that day was Ju 88 A, W.Nr. 3036, 5J+HT of 9./KG 4, which went down while attacking AAA positions at Le Havre. The bomber's crew (wounded Ofw. W. Hansen, as well as Uffz. G. Lewandowski, Uffz. H. Krahn and Gefr. K. Rauchfuß) were captured by Allied troops.

On the same day Ju 88 A-1 z 2./KG 51 crashed on take-off from its base at Böblingen, killing Uffz. S. Stretz, Uffz. J. Putzbach and Uffz. H. Böttcher. Another Ju 88 A-1, this time from 3./KG 51 was jumped north of Braine by three Potez 631 fighters from ECN 1/13 and 4/13 and shot down before reaching its target. The machine crashed near Nampcel killing Fw. M. Pfeiffer and Uffz. K. Lang. Critically wounded Oblt. W. Schallenberg managed to bail out of the doomed bomber, but he died later in hospital at Soissons. Fw. M. Roggentin had much more luck: he survived the fighter attack and was captured by the Allies.

As the bombers from I. and II./KG 51 were dropping bombs on Cateau-Thierry train station, they were attacked by the RAF fighters which knocked down three of them. Ju 88 A-1 from 5./KG 30 was first engaged by AAA and received a direct hit in one of the engines, then finished off by a Hurricane from No. 73 Sqn. The bomber's crew (Oblt. V. Schnez, Ofw. H. Ullrich, Uffz. O. Runge and Ogefr. H. Muhle) bailed out and were all captured. Ju 88 A-1 from 5./KG 51 was shot down by F/L Scoular flying a Hurricane from No. 73 Sqn. and crashed near Courville. Two of the crew member were killed (Lt. F. Thelen and Fw. H. Badura), while Gefr. H. Blicker and Uffz. R. Rüthrich ended up as POWs. Another Hurricane from No. 73 Sqn. (flown by P/O Paul) dispatched a Ju 88 A-1 from 4./KG 51. The German crew survived, but were forced to crash land their machine near Karlsruhe. Fw. Friedrich Aufdemkamp from II./KG 51 had a rather terrifying "adventure" when pieces of engine cowling separated during dive recovery and damaged the bomber's port horizontal stabilizer. The German pilot managed to recover from the dive and brought his machine in for an uneventful landing at his home base.

On May 20, 1940 the Wehrmacht units continued their push towards the Channel. Amiens fell at noon and by the end of the day the German tanks of 2nd Panzer Division reached the coast at Noyelles, near Abbeville. The Allied troops still fighting in Belgium and northern France were trapped.

On that day Ju 88 A z 4.(F)/122 crashed near Dülmen, just west of Münster during a forced

Ju 88 aircrews were very busy over Dunkirk sinking numerous Allied vessels. Pictured here is the sinking French destroyer *Bourrasque*.

This Ju 88 A-1 was shot down by the RAF fighters over Dunkirk.

landing attempt. Oblt. E. Gaul sustained serious injuries, but the rest of the crew (Lt. H. Doermer, Fw. F. Bach and Ogefr. K. Lessmöllmann) survived the crash with only minor scratches. Another Ju 88 lost that day was Ju 88 A, W.Nr. 3051, 5J+IR, which was shot down over the sea near Zeerbrugge by French fighters from GC I/4. Oblt. H. Camphausen, Uffz. R. Gäbeler and Ofw. R. Rieck were all listed as killed in action. At 09.35 French fighters of GC III/1 downed near Compiegne a Ju 88 A from 3./KG 51. Gefr. E. Koschinski, Fw. A. Frühling, Uffz. H. Rinn and Flg. W. Gilgen bailed out safely and were captured by the Allies.

Later that day Ju 88s from KG 30 bombed a Belgian transport *Antverpia* (4,932 tons), whose crew was forced to beach her on the following day.

In the late evening Oblt. Hajo Herrmann from 7./KG 4 flew his first combat sortie over Dunkirk: "Our attacks on sea convoys around Ostend, Dunkirk and Calais were coming fast and furious. Dunkirk got hit especially hard on May 20 when the British were trying to evacuate their troops. On that day, at around 23.00, I saw a large freighter bathed in bright moonlight, which was just steaming out of Dunkirk and heading north. I didn't attack immediately because I didn't want to get caught in British searchlights and find myself at the receiving end of heavy flak guns. After about half an hour and constantly keeping an eye on the traffic below, I saw the freighter heading out along the shore on the easterly course, most likely to stay within the mine-free corridor before she could set course for Dover.

I finally made a decision to attack and chose the profile that I had learned from Storp back in Spain: I wanted to hit four different ships using the four 250 kg bombs I was carrying. And there was a multitude of targets to chose from. As I was getting ready another transport was leaving the peer.

I set up my attack along the ship's axis coming in with the moon behind me. I could see the target clearly. I armed the bomb, went into a dive and pressed the pickle button. When I recovered from the dive the gunner reported: '10 meters short.'

I climbed up again and set up for another pass: 20 meters long. My observer, a rookie Feldwebel, a good guy but a bit of a pessimist, immediately suggested dropping the two remaining bombs in a salvo. This way, he said, we would have something to show for ourselves. As I was climbing back up for another pass, I came close to actually following his advice, but the Feldwebel kept whining and moaning, which made me really angry, so I said: 'No, I'm dropping just one of them.' I knew I was taking a gamble, but the temptation was too hard to resist. There was a flipside to that madness, too. On the one hand, I had already accumulated quite a lot of experience since my time in Spain. I was probably more experienced now then Strop and I put together back then. Second, thanks to the first to drops I had the target bracketed really well, so it was not unreasonable to expect a direct hit on the third try.

I dived in and placed the bomb right on target. I could see it almost immediately as there was a flash of bright light shooting skywards.

When we were climbing up to altitude I saw a small fire erupt on the deck, which made me wonder if one more bomb was perhaps in order to finish the ship off. However, within the next ten minutes the fire engulfed the entire ship. The transport was lost. I could see a number of corvettes and other ships rushing to the scene to pick up the survivors. I could only imagine hundreds of troops jumping off the ship to save their lives.

I later estimated the ship's tonnage at 8,000 tons, but other estimates were as high as 10,000 tons."[29]

Arrival of the Wehrmacht units on the Channel coast on May 21, 1940 reinforced the flanks of the German advance. The Luftwaffe ruled the skies above the battlefield, but the Allied fighters still posed some danger to the Luftwaffe bombers. Soon after 12.00 a Ju 88 A belonging to 3.(F)/121 was attacked south of Amiens by a flight of fighters from GC II/1, flown by Adj. Richardin, Sgt. Roquerbe and por. Marian Trzebiński. The attack put the bomber's starboard engine out of action and when another pair of French fighters from GC II/9 joined the fight, the Luftwaffe crew were forced to bail out of their wounded bird. The Junkers crashed at 12.30 near Remy, taking with it Uffz. R. Bomsdorff, who had been killed in the initial attack. Lt. H. Wittmeyer and Ogefr. W. Sauer did manage to bail out, but did not survive the jump. Only Lt. O. Nitschke reached the ground safely despite being shot in the shoulder and was soon captured by the Allies.

Another Ju 88 A from 4.(F)/121 made an emergency landing near Saargemünd following an electrics failure. The bomber was completely consumed by fire, but the crew survived without a scratch. Also 9./LG 1 lost a Ju 88 A-1 that day. The machine was forced down by a fighter from No. 54 Sqn. flown by F/L Leathartand and crash landed just west of Boulogne. Gefr. H. Rose, critically wounded in the attack, was taken to hospital at Cambrai, but the other crew members (Fw. F.-K. Meinshausen, Uffz. G. Jakubowski and Uffz. R. Tilser) suffered only minor injuries.

In Dieppe the bombers of KG 30 sank a small steamboat *Hubbastone* (873 tons), a passenger ship *Maid of Kent* (2,963 tons) and four Belgian trawlers. Another attack By Ju 88s resulted in the sinking of a Panamanian tanker *Clairy* (5,838 tons), which went to the bottom 9 miles from Boulogne. The entire crew survived the attack and was picked up by friendly vessels.

May 22, 1940 was a busy day for Ju 88 crews from KG 30 operating over Dunkirk, although they failed to achieve any spectacular success. Two of the wing's machines were lost that day to enemy action. The first one, Ju 88 A-1 from 3./KG 30 was shot down by a Spitfire of No. 74 Sqn. The bomber's crew (Uffz. W. Mattner, Uffz. E. Klose, Uffz. W. Gottlebe and Gefr. J. Corsepius) were all subsequently listed as missing. A Hurricane pilot from No. 151 Sqn. claimed Ju 88 A-1 form 8./KG 30, which crash landed south of Calais. Uffz. L. Georgi, Uffz. K. Spura and Gefr. P. Wieczorek survived the crash with only minor injuries.

On the same day III./LG 1 lost three Junkers bombers in operations over Boulogne. Ju 88 A-1, W.Nr. 3025, L1+DS was downed by a Spitfire from No. 54 Sqn. Gefr. Pohle was killed in the attack, while Oblt. H. Dunse, Fw. H. Kakanowski and Gefr. H. Przybyla were listed as missing. Another Ju 88 A-1 from 8./LG 1 was shot down during the attack on the tanker *Clairy* by Spitfires from No. 65 Sqn. The British fighters heavily damaged the aircraft's starboard engine, forcing the crew to crash land their machine near Montreuil. The Junkers was completely consumed by post-crash fire, but Gefr. G. Glogger, a Oblt. T. Hagen, Ofw. K.-R. Britting and Uffz. H. Döring escaped unscathed. Ju 88 A-1 z 9./LG 1

Make shift peers in Dunkirk were built using wrecked trucks.

sustained heavy damage in an encounter with a Spitfire and crashed while the crew attempted an emergency landing in Düsseldorf. Fw. H. Jahn suffered serious injuries, while Oblt. K. H. Schomann was only slightly injured.

Over the next three days, despite a very high tempo of operations, the Luftwaffe Ju 88 units suffered no combat losses. No Ju 88s were lost until May 26, i.e. a day before the evacuation of the British Expeditionary Force began. At 06.00 Ju 88 A-1, W.Nr. 4014 of 9./LG 1 went down while attacking Allied shipping near Calais. Following the attack by British fighters the bomber caught fire and crashed just north west of St. Omer, killing Ofw. F. Körnig, Uffz. R. Kuke, Gefr. C.-H. Krüger and Gefr. H. Steinhagen. During an afternoon strike on Dunkirk KG 4 lost two of its Ju 88s. The first one was Ju 88 A-1, W.Nr. 3049, 5J+KR of 7./KG 4, which crashed into the sea at 14.15 following an attack by P/O Cooper-Slipper from No. 605 Sqn. The bomber's crew (Fw. F. Weigel, Hptm. Dr. H.-J. Rabe, Ogefr. J. Koch and Gefr. G. Thiel) were all listed as killed in action. The second machine that did not return to base that day was Ju 88 A-1 from 9./KG 4 shot down at the same time by an RAF fighter from No. 32 Sqn. flown by F/L Crossley. The Luftwaffe crew (Oblt. F. Podbielski, Fw. H. Berkefeld, Uffz. K. von Kiedrowski and Uffz. O. Piontek) bailed out safely. Ju 88 A of 8./KG 30 crashed shortly after take-off from Schiphol as the crew were heading out to attack targets around Dunkirk. The huge explosion claimed the lives of all on board: Uffz. G. Dujardin, Uffz. W. Bethe, Ogefr. L. Burk and Gefr. J. Lippitsch.

The massive effort to evacuate British troops from Dunkirk began on Monday, May 27, 1940. That day Ju 88s from KG 51 were on scene to disrupt the British evacuation. Four of the wing's machines returned to base with various degrees of battle damage and two were lost to enemy action. Ju 88 A from I./KG 51 crashed near Mannheim while the crew attempted an emergency landing following an encounter with AAA over Dunkirk. The aircraft was a total write-off, but all crew members survived with no injuries. Ju 88 A of 2./KG 51 was badly damaged by AAA and British fighters and crashed on landing at its base at Brussels-Evere. Lt. A. Berger suffered serious injuries in the crash, while Uffz. H. Puls was only slightly injured.

KG 30 insignia painted under the canopy of a Ju 88 A-1.

II./LG 1 lost two aircraft during a sortie over Ostend. The first one was a Ju 88 A-1, which crashed on landing at Düsseldorf-Lohausen following an attack by enemy fighters. The CO of 4./LG 1, Hptm. H.-D. von Kobylinski, sustained injuries in the crash. The second machine (L1+KP) was shot down near Ostend. The crew bailed out and escape injuries, except Uffz. G. Kunz, who was slightly wounded in the fight.

The only loss that the Ju 88 units suffered on May 28 was the Ju 88 A of 8./KG 30, which was damaged by AAA while on a reconnaissance sortie over the Channel. The aircraft crashed into the sea killing Uffz. K. Geuthner and Gefr. G. Follrichs. The other two crew members survived and managed to swim back to the shore.

The weather over Dunkirk improved greatly on May 29, which allowed the Luftwaffe to increase the tempo of operations. The German bombers sank a destroyer (HMS *Grenade*), a minelayer (HMS *Waverly*) and at least fifteen merchant vessels. In addition, three destroyers were badly damaged by German bombs. LG 1 lost four aircraft over Dunkirk and one of the unit's bombers was badly damaged. Ju 88 A-1 z 5./LG 1 crashed near Mettmann due to battle damage suffered in an air-to-air fight, killing Ofw. A. Sasse and Fw. H. Ruppel. Uffz. H. Roewer and Uffz. E. Berwanger suffered only minor injuries. Ju 88 A-1, L1+AP flown by the CO of 6./LG 1 was shot down by Hurricanes from No. 151 Sqn. and crash landed on the beach. Ofw. F. Stobbe and Flg. S. Teßmann were killed in action, but Staffelkapitän Oblt. A. von Oelhaven and Fw. W. Notzke survived the crash and were captured by the Allies. A direct AAA hit destroyed the starboard engine of Ju 88 A-1, L1+LP from 6./LG 1, which forced the crew to make an emergency landing at Terhofstedepolder. Gefr. K. Meier sustained minor injuries to his left hand, but Ofw. Hahne, Fw. H. Gießen and Uffz. Bronitzky walked away from the crash unscathed. The bomber was completely destroyed by post-crash fire. Ju 88 A-1 z 6./LG 1 came under attack by five Defiant fighters from No. 264 Sqn. and crashed near Ostend killing Flg. H. Oehler. The other crew members (Fw. M. Stapelberg, Uffz. H. Rosenow and Gefr. H.-J. Brömel) suffered serious injuries and were taken to hospital. Another LG 1 bomber (Ju 88 A-1 of 9./LG 1) was jumped by Skua fighters from No. 806 Sqn., but its crew managed to nurse the damaged machine back to base. Oblt. H. Becker and Uffz. R. Tilsner were wounded during the battle.

Another photograph of a makeshift peer built on the beach at Dunkirk.

On the following day the Luftwaffe operations over Dunkirk were once again scaled down due to bad weather. The only Ju 88 loss that day was the A model of 1./KG 30, which was most probably shot down by a Coastal Command aircraft over the North Sea.

By the afternoon of May 31 the weather had improved and the Luftwaffe bombers immediately returned to the skies over Dunkirk. The bombs dropped by Ju 88s from KG 4 finished off the French destroyer *Sirocco*[30], which had been previously damaged by a German torpedo boat. Also damaged by the bombs were British destroyers HMS *Express*, HMS *Icarus*, HMS *Impulsive* and HMS *Scimitar*, as well as the French destroyer *Leopart*.

Oblt. Hajo Herrmann from 7./KG 4 was among the Ju 88 aircrews operating over Dunkirk that day: "We launched in a large combat formation from Schiphol near Amsterdam and headed out across Belgium to attack besieged Dunkirk, especially ships, boats and ground troops amassed on the beach and along make-shift peers. I was flying a spare machine, which I hadn't been able to properly pre-flight, since my aircraft had a malfunction shortly before take-off.

Even from very far away it was impossible to miss Dunkirk as its location was marked by pillars of black smoke rising high above burning oil tanks in the harbor that mixed with a thin layer of clouds above the city. I can still remember that sight.

Our escort fighters and destroyer aircraft cleared the sky ahead of us of all enemy fighters, which made it easy for my flight to maintain proper position in the formation.

We were approaching our diving point amongst bursting AAA shells. The aircraft ahead of us were diving in on their targets, only to pull up again just a short moment later. Now was my turn to reach for the handle and extend the dive brakes, but when I pulled the lever, nothing happened. I was going downhill at a tremendous speed passing other aircraft along the way. I could hear the rush of wind outside the cockpit rise to a roar as the entire aircraft began to shake. I pulled the dive brake lever again – nothing! Well, that madness had to stop. I recovered from a dive as gently as I could, careful not to lose the wings. Other aircraft around me were all swooping down to drop their bombs. In the meantime, my crew and I were the only ones perched up high in the sky. Let's join the rest, I thought, let's give it one more try. There were so many boats to choose from down below!

Suddenly a silhouette of a Hurricane whizzed right past us, then another. Up, up, into the safety of the clouds! Within minutes we were in the clouds, but then... we popped up on top almost as quickly! So down we went again and then up once more. That was one thin cloud layer! Suddenly the cockpit was peppered with machine gun bullets exposing a bunch of wires behind the instrument panel. Should I jettison the bombs? No way! Both engines were still turning, weren't they? So back into the clouds we went. I was careful to keep an eye on the cloud tops hoping not to pop out above them too soon. This time a different, strange-looking aircraft was waiting there for us. Could it be a French Dewoitine? I was sure I saw something similar to that back in Spain, a very agile beast. But our bird was nimble, too, so we went back down into the clouds, but not too deep. Then there was a tremendous bang. Oh boy, the port engine was

Junkers Ju 88 A-1 of KG 51, which fell victim to the British fighters.

out. The radio operator was screaming, the gunner moaning. Behind us there was a thick trail of dark-brown smoke. I pushed the nose down and hurtled towards the deck at full speed. The fighters broke off, either because they ran out of ammo, or because they were gallant chaps. What do I do next? Land among the Tommys[31] on the beach? Absolutely not! The thought 'whatever!' went briefly through my mind and then I pushed the machine even closer to the deck and made a wide turn to the left, then to the right, heading straight for a medium-sized transport at the peer. I then released the bombs from just 10 meters, but they all missed the mark.

I push the throttle of the single good engine to get the hell out of Dodge as soon as possible. At that moment one angry Hurricane pilot pops out of nowhere and there is another deafening bang, but nothing much happens. The guy was alone and then he seriously overshot us, so there was no way he could do us any serious harm. Now every second was precious. I kept on heading east trying to get as far away from the beach as I possibly could. I could still hear burst of machine gun fire hitting the water. None of us was shooting. The guys behind took cover and I tried to fit my entire body behind the armor-plated back of my seat. Then we took a hit that finished us off. There was a tremendous crashing sound, but I couldn't even tell where it came from. The water rushed in through the shuttered windshield, taking my breath away for a brief moment. My moaning Feldwebel jettisoned the canopy in the last possible moment and I was under water. What am I waiting for? Let's get out of here!

I tore off every piece of gear I was wearing: the life vest, flying suit but when I climbed out of the cockpit... I stood up, chest-deep in water, standing on the fuselage. Before long three more heads popped up. The incessant sound of big guns was coming from the beach.

The two of us up were in one piece, but the guys in the back had some nasty cuts and lacerations on their hands and faces. We all slid down the fuselage and started towards the shore. After a tough few minutes fighting the swells we finally reached dry land and took a very careful look around. Suddenly, a German helmet popped up from behind the tall grass. Buddy, let me give you a big hug!

We were saved, we made it. It was my 40th combat mission of the war, on top of over a hundred sorties I had flown in Spain."[32]

Oblt. Hajo Herrmann's aircraft was most likely shot down by Sgt. Robinson flying one of the Hurricanes of No. 111 Sqn. KG 4 lost one more bombers in the same operation, Ju 88 A-1 from 8./KG 4, which was badly damaged by Allied fighters and AAA. The crew did manage to bring the machine back to base, but the damage was assessed to be beyond repair and the aircraft was subsequently written-off. One of the crew members, Oblt. D. Marwitz, sustained serious wounds and was taken to hospital in Amsterdam.

On June 1, 1940 both bomber and reconnaissance versions of the Ju 88 operated over Dunkirk. One of the latter (Ju 88 A, F6+DK z 2.(F)/122) was lost at around 10.00 when it came under attack by F/O Moberley from No. 616 Sqn. Lt. R. Hey, Fw. E. Lange-Gläscher, Fw. E. Kieß and Uffz. W. Giesa were all captured. Another Ju 88 A, this time from 3.(F)/123, had a close encounter with the British fighters of No. 151 Sqn. and French machines from GC II/8, but its crew made it back to base safely. Two more Ju 88s were lost that day over central France. One of them was Ju 88 A, W.Nr. 8004, 9K+DM, of 4./KG 51, which was shot down at 13,45 by a French fighter from GC III/9 over Ruffieuex. The bomber's crew (Fw. R. Müller, Uffz. P. Dickhut, Gefr. B. Sartor and Gefr. H. Spieß) survived and were taken prisoner. The other aircraft was Ju 88 A, 7A+PM, belonging to 4.(F)/121, which fell victim to French fighters from GC II/2 and GC II/7 while on a reconnaissance mission over Besancon – Dijon. The bomber crashed near Marais du Goutterot, but the crew survived and ended up as POWs (Oblt. H.-W. von Niebelschütz, Ofw. E. Winkler, Uffz. H. Maiwald and Uffz. A. Dobschall).

June 2, 1940 was the final day of the Luftwaffe's operations over Dunkirk, which was finally captured by the Wehrmacht on June 4, 1940. When the first German troops arrived at the beach strewn with abandoned equipment and wreckage of all kind, they stumbled upon a sore-looking Luftwaffe flyer. The man was Oblt. von Oelhaven, the CO of 6./LG 1, who was captured by the British after his Ju 88 had been shot down on May 29. As Oelhaven was being led along the make-shift peer built from truck wrecks to board a British transport on June 2, he jumped into the water and hid in darkness among wrecks and floating debris, where he remained for 36 hours!

On the same day 3./KG 51 lost another Ju 88 A over France. The bomber was shot down by fighters from GC III/9 near Dagneux, north east of Lyon and crashed killing the entire crew (Oblt. P. Kibele, Ofw. M. Schulzki, Fw. G. Köppel and Uffz. R. Bauer).

June 3, 1940 saw the launch of the largest Luftwaffe operation to date, code-named "Paula". It's main objective was neutralization of the remnants of French air force based at airfields around Paris and bombardment of aircraft manufacturing plants located in the vicinity of the nation's capital. The operation involved around 1,000 aircraft, including Ju 88 bombers.

The offensive began at 14.00 with a coordinated attack on 13 French airfields. However, the strike did not produce the expected results, since the French launched their fighters before the Luftwaffe onslaught began. Only 16 French aircraft were destroyed on the ground by German bombs, another 47 sustained damage and 28 were lost in air-to-air engagements. Of the fifteen aircraft manufacturing plants on the target list, only three suffered significant damage.

The Luftwaffe lost two Ju 88 As that day. One of them was a machine belonging to 8./KG 4, which hit a barrage balloon cable during a nighttime attack on Le Havre. The crew managed to bring the damaged bomber home back

The French air base at Romorantin under attack by Junkers Ju 88s of KG 51.

to Schiphol, but it didn't survive a crash landing attempt there. Staffelkapitän Hptm. W. Storp sustained injuries during the crash. Another bomber lost that day was 9K+AA from Stab KG 51 flown by Geschwaderkommodore Obst. Josef Kammhuber. The crew were attacking a French airfield at Etampes when their bomber was jumped by French fighters and crash landed near Cocherel at around 14.00. The entire crew (Obst. J. Kammhuber. Lt. H. Kroesch, Lt. W. Lüderitz and Uffz. K. Schneider) were captured.

While the Allies suffered a crashing defeat in Belgium and Flanders, the remaining French forces were still strong enough to continue combat operations. The French army consisted of at least 78 divisions, in addition to three British, three Polish and a single Czech division. The total force of about 85 divisions was supported by 1,285 French aircraft, of which 804 were fully combat-ready. What made the real difference though was the low morale of the French troops.

Facing the Allies were 142 German divisions of well-trained and highly motivated troops who fought with unwavering belief in the Reich's final victory. In the air, some 1,500 fully combat ready Luftwaffe aircraft ruled the skies.

Operation "Fall Rot", the final and decisive stage of the French campaign, began at dawn on June 5, 1940 with massive attack of Heeresgruppe B along the front between the Channel and the river Aisne. The Germans launched their offensive from a few bridgeheads on the Somme using a solid air support. During a strike on Toulouse II./KG 51 lost three Ju 88 A-1s. Ju 88 A-1, W.Nr. 29, 9K+HM was shot

Junkers Jumo 211 B-1 undergoing maintenance.

The B-Stand in a Ju 88 A-1. Clearly visible are two additional flexible MG 15 machine guns mounted in the canopy's side windows.

down by French fighters from GC I/2 at 13.45, just south of Troyes. Lt. H.J. Zobel, Oblt. P. Herr, Fw. E. Tocha and Ofw. K. Graffenberger became POWs. At 13.50 Ju 88 A-1, W.Nr. 5018, 9K+KM fell victim to the aircrews of GC I/2 over Recey-sur-Ource and crashed during a forced landing attempt near Arbot. Oblt. F. Hohenstein and Fw. R. Hepp perished in the crash, while Fw. H. Kroll (wounded) and Fw. H. Ludwig were captured. Ju 88 A-1 from 5./KG 51 was shot down by fighters of GC I/2, killing Uffz. A. Schopf. Uffz. E. Werner managed to bail out of the bomber, but his parachute harness failed just as the canopy was deploying, which sent him falling to his death near Chamignol. Oblt. H. Kohl survived the jump, but suffered serious leg injuries, while Fw. Gustav Doege landed safely, but was murdered on the ground by French infantrymen. Commandant Reverseau, CO of 3rd Battalion, 108th Infantry Division explained later that rear gunner Gustav Doege acted aggressively and was shot dead by a French officer, who feared the German was trying to reach for his weapon as he put his hand in the pocket.

Another Ju 88 A, this time from Stab I./KG 51, was shot down by French fighters near Romorantin-Prunieres at 14.00. The bomber made a successful forced landing near l'Angelus where the entire crew (Fw. K. Diekert, Fw. H. Wagner, Uffz. W. Krichbaum and Gefr. W. Schindler) were captured.

Ju 88 A from 2.(F)/122 hit an obstacle while taking-off from Köln-Ostheim and crashed killing the entire crew (Lt. J. Nolte, Uffz. B. Renken, Ogefr. K.-H. Rott and Gefr. L. Hase).

At around 18.30 III./LG 1 lost two aircraft attacking railroad targets near Rouen. The bombers came under attack by fighters from the RAF's No. 501 Sqn. and French machines of GC II/10. The crew of Ju 88 A-1 from Stab III./LG 1 survived the attack and ended up as POWs (Uffz. G. Weimann, Ofw. J. Tiedemann, Fw. K. Brammann and Ogefr. K. Franz). Also captured were Fw. J. Jankowski, Uffz. K. Zipperling, Ogefr. G. Sirlin and Gefr. H. Plate, who bailed out of Ju 88 A-1 from 7./LG 1.

On June 6, 1940 German ground forces broke through the French defenses in several spots along the front. The Wehrmacht was beginning to take control of the situation. By that time the Ju 88 units had been almost entirely withdrawn from operations, which gave their commanders time to regroup and devote more time to operational training.

On the following day the tanks of XV. Armeekorps broke through the frontline and pushed to within 36 kilometers of the Seine and Rouen. On Saturday, June 8 the entire French defensive line along the Somme collapsed. The defenders began their retreat towards the Seine, where new lines of defense were hoped to be established. June 9 brought more bad news for the French army. On that day Armeegruppe A launched a massive offensive along 150 km frontline., stretching from the Seine all the way down to the Ardennes. The German's main objective was to capture bridgeheads on the Aisne, which would allow the Wehrmacht to continue their drive towards Paris. In the meantime Armeegruppe B captured Compiegne and Rouen. On the following day the continuing German offensive punched through the French defenses and armored spearheads were closing in on Paris. On the same day Benito Mussolini declared war on France and Great Britain. On June 11, 1940 when Paris was declared an

This Junkers Ju 88 A-1 from I./KG 51 crashed during an emergency landing attempt following an engine failure.

"open city", the French forces were still trying to hold the line along the Marne. In the meantime, the German attack continued across the Aisne and by the end of the day the Wehrmacht troops reached Reims. Over the next two days bad weather grounded most of the Luftwaffe aircraft. At 05.30 on June 14, 1940 the units of the Wehrmacht 9th Infantry Division rolled into Paris. June 15 saw the first Ju 88 losses in ten days. Ju 88 A-1, W.Nr. 004, 5J+GS from 8./KG 4 was shot down by S/Lt. Thollon from GC I/8 while attacking bridges on the Loire, just east of Orleans. The bomber was attacked over Jargeau and crashed just outside Fay-aux-Loges at 18.15, killing Lt. J. Reiners, Uffz. R. Haase, Uffz. W. Püschel and Uffz. W. Thier. The crew of Ju 88 A-1 from 9./KG 4 had more luck and survived a crash landing after a sortie over Paris.

In the meantime the crews of II./KG 30 were tasked with bombing port installations at Cherbourg. One of the pilots involved in the operation was Uffz. Peter Wolfgang Stahl: "At 15.00 we received operational orders to launch an evening raid on Cherbourg. A seven-ship formation was to be led by Hptm. Peters. The crew from our flight was Fw. Kühnast, Fw. Lorbeer, Fw. Erkens and Uffz. Stahl.

We went into the briefing, where each crew were assigned individual targets. My crew was to perform a diving attack on warehouses within the port. At 16.30 we launched for a hop to Amiens with only half-full fuel tanks, since with full fuel load we wouldn't have been able to launch from that particular field. The plan was to refuel at Amiens before launching again for the long flight to target area.

The take-off from the makeshift, soggy strip, even at low weight, was a terrifying experience, almost life-threatening. I guess for a combat flyer the war starts even before he can face the enemy eye-to-eye. At Amiens there was still a lot of evidence of the presence of our English "friends", who had been tenants there until not that long before. Jupp Harras, who was always on the lookout for goodies, even found his first war booty – a crate full of English candy!

At 19.00 sharp Hptmn. Peters launched from the dusty strip and I soon followed. I had to roll completely blind through a cloud of dust kicked up by my leader and I couldn't be happier when I felt my heavy machine finally got airborne. My hands ran the drill moving through different levers in the cockpit: gear, flaps, fuel pumps, prop pitch, radiator flaps. My bombardier raised his left hand signaling all was in order on his side as well. Scanning the instruments, I was slowly closing in on Hptmn. Peters. Others were forming up behind us and soon we all turned west. Each machine carried 3,600 liters of fuel, more than enough to complete the sortie.

Initially we stayed fairly low and could see bomb-scarred towns and villages passing down below. There was the unfinished Weygand Line – a system of fortifications designed to stop the German attack. From the air we could see lots of abandoned artillery pieces and vehicles. The roads, crowded with refugees, presented a very sad sight. The fields down below were littered with dead cows, which died because there was nobody there to milk them. That grim and dreadful landscape made me forget for a minute that we too were heading out to be part of that hell.

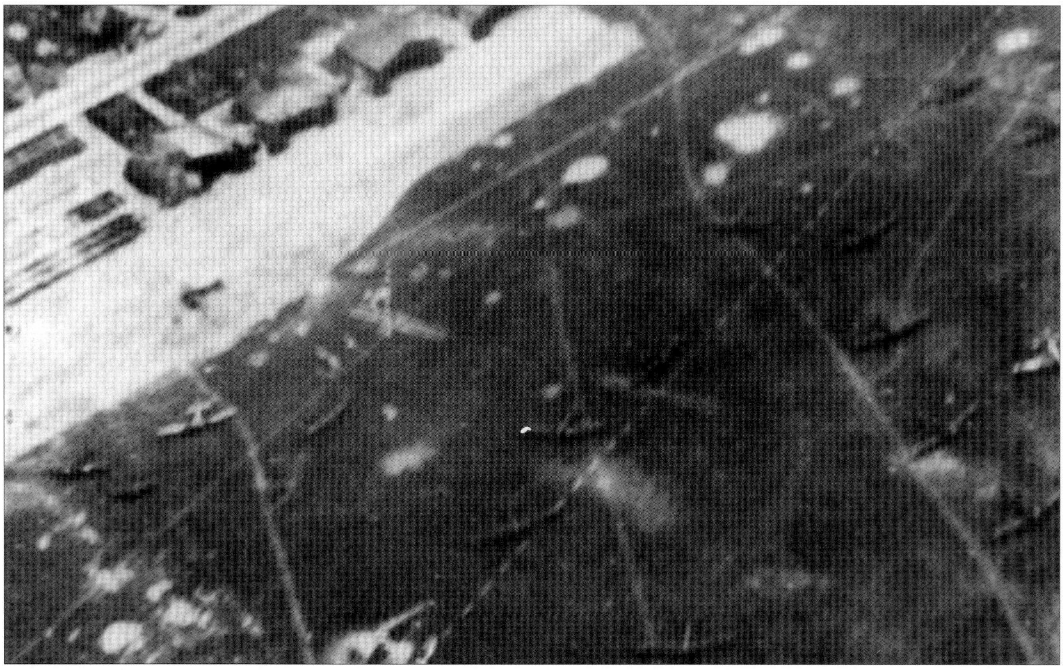

Paris-Le Bourget airfield being bombed by Ju 88s from KG 4. Operation "Paula", June 3, 1940.

West of Le Havre we went feet wet to reach the target area from the sea, hoping to stay out of range of the enemy flak. Nobody knew exactly where the frontline was. There were still fires burning in several neighborhoods of Le Havre, despite the fact that the city had been under our control for the past few days. The port was strewn with all kinds of ship wrecks.

Flying over the Channel we could see the French coast in good detail on the left and the barely visible British coast on the right.

We leveled off at 4,000 meters and after a pre-determined period of time turned due south keeping our eyes peeled looking for Cherbourg.

I was flying slightly above the leader to have a good view of the attack before I commenced my own bomb run. After a few minutes we were visual with the target dead ahead. The machines in front of me dived almost simultaneously. At almost the same time I acquired my target. I took one more glance outside. It was the first time I was dealing with AAA trying to bracket my own aircraft. I felt the flak posed no danger, since the dark puffs of exploding shells were popping up at least 50 meters away. Besides, the AAA was not too intense. With complete calmness I positioned the target in the glass panel fitted in the floor and was ready to commence my dive. At the same time I ran through the cockpit flow required before the dive: prop pitch, auxiliary fuel pump on, radiator flaps closed, altimeter set and audio warning signal armed for 800 meters. Jupp armed the bombs and set the ripple release system. We all checked our seat harnesses. Then I set the trim wheels for a dive and we were good to go!

In the last moment (I almost forgot) I turned on the reflector sight and set the ring and deflection. I then pulled back on the throttles almost to flight idle and pulled the lever actuating the dive brakes, which caused my Ju 88 to almost roll inverted and go into a dive. We were hurtling downhill. I kept the target, looming ever larger, in the ring. Gentle control inputs kept the target centered in my sight. One last glance at the horizon, which by that time was not in front of us, but above, visible through the roof of the canopy. The flak didn't bother us too much, or at least I didn't think it was an immediate threat even as we were flying through some of the black puffs with the smell of cordite entering the cockpit.

The dive was going as smoothly as in training when a signal in the headphones announced the release height. Now I had to keep the machine perfectly steady, or the bombs would miss the mark. I was dropping like s stone when I finally pushed the red button on the control column. The aircraft shuddered – bombs away! The aircraft went into automatic recovery mode. Four maybe five g's pushed us hard into our seats and threw back our heads, which until the moment of release were all hung low almost on our chests. I blacked-out briefly during the recovery. We used to call it a "grey curtain". Seconds later we were all fully conscious again. We were going straight up with the air speed indicator still showing well above 500 km/h!

Now was the time to fly some evasive maneuvers, because we were in range of medium flak guns. I could sea their red tracers crisscross the sky ahead of us. From his battle station in the ventral tub, Moritz could observe our bombs fall and was now confirming good hits.

Even though I couldn't see my comrades in other bombers, I knew exactly where they were: mad AAA fire of all calibers was pointing in their direction. I liked that, because I could safely egress the target area even if it meant flying crazy evasive maneuvers. Suddenly, it was dead quiet and I could focus on the return leg back to base. (...)

We were the last crew to land at almost completely dark airfield at Le Coulot. Only very few machines returned to base unscathed, there were even two badly wounded among the crews. The old hands claimed the AAA was heavy or very heavy over the target, which I was obviously unable to assess myself. All I knew was that the tactics I chose seemed to work pretty well."[33]

On the following day, June 16, 1940 II./KG 30 was tasked with bombing a bridge on the Loire at Tours. The sortie was flown by crews commanded by Oblt. Stoffregen, Lt. Mann, Uffz. Peter Wolfgang Stahl, Fw. Erkens and Uffz. Schröder. The bombers took off at 12.45 in decent weather with 4/10 cloud coverage and tops at 3,000m. After two hours the formation arrived over target area, which was heavily defended by AAA. Uffz. Stahl remembers: "Our flight was the first one to arrive over the target. Stoffregen made a wide turn to the west in order to come in on his bomb run from the east. Erkens and Schröder turned east before extending far out to the south. This resulted in a split in AAA fire, which was very helpful to us.

I decided to repeat my tactics from the day before – I positioned behind the lead and to the right. This way I had a good view of the action and was able to dive in on the target with the sun behind my back.

I saw the two aircraft ahead of me pitch up almost simultaneously to begin their dives. The entire way down to the target they were followed by the puffs of bursting flak shells. In the meantime, I wasn't bothered by anyone. I was just hoping the flak gunners wouldn't get me as I followed behind my colleagues.

As soon as I put the aircraft on the attack course I saw Stoffregen and Mann's bombs explode down below and miss by just a little bit. I began the same, calm dance I did the day before: fly straight and level without unnecessary acceleration, configure the aircraft for the dive, arm the bombs and approach to the diving point. Once the target passed the red line painted on the floor window I pulled the dive brakes lever. The nose dropped immediately and I took a quick scan of the airspace around us. They hadn't spotted us yet. As I was entering the dive, I noticed that the flak fire had shifted to the south, which was where Erkens and Schröder where coming in from. I then focused entirely on the dive and keeping the target centered in my sights. Suddenly I saw AAA shells exploding right in front of me, along my dive path. Then Groß screamed: 'We're hit!' I instinctively hit the auto recovery button on the left side of the cockpit, which took us out of the dive without the bombs release. There was a brief 'grey curtain' moment, followed by the view of a blue sky. Is everybody ok..? Yes! What happened? The engines were running smoothly, all instruments were working as advertised, but one of the canopy glass panels was shattered causing a very disturbing hissing and whining noise of the rushing wind. Apart from that, there seemed to be nothing wrong with the aircraft. I immediately made a decision to try again. But first we had to climb back to altitude. I pushed the throttles forward using some of the momentum of the recovery to speed up the climb, but it still took what seemed to be ages. I saw Erkens and Schröder come in from the south to commence their bomb runs. They were under tremendous pressure from AAA. In the meantime I continued to climb into the sun, unmolested by flak. 'Look out for fighters guys!', I said. When Erkens and Schröder had recovered from their dives, I reached 2,000 meters in altitude. I knew I had to attack immedi-

This Ju 88 A of 4.(F)/121 crashed at one of the airfields in France.

Junkers Ju 88 A-1 sitting under camouflage netting.

ately, otherwise a single straggler like me would be an easy target for the flak gunners.

I turned towards the target when Erkens and Schröder were still in a dive. I was coming in as planned, from the sun, and nobody was shooting at me. I went through the motions one more time: I saw my friends' bombs explode below, kept the target in sight and went into a dive. It all went like a drill. I waited until just below the release height before dropping the bombs, recovered, the 'grey curtain' fell as usual, then there was a blue sky. Let's get out of here!

I squeezed every last bit of power from my Ju 88 and heard Moritz report good hits. AAA opened up on us, but I managed to steer clear of the bursting shells. Then it was all quiet again. I leveled off and set course for the home plate. The weather got worse and we flew over Paris in a thunderstorm. After a while we were flying over the old battlefields of the Great War. When we finally taxied in at Le Coulot at around 17.10, everybody else had already landed."[34]

5./KG 30 lost one of its bombers that day, which was first badly damaged by AAA and then finished off by fighters from DAT Tours. The Ju 88 A-1 crashed near Flee, killing Fw. K. Hacker, Uffz. F. Richter and Flg. D. Fleßner. Badly wounded Lt. W. Mann was taken to hospital in Cologne.

On June 17 84-year-old Marshal Petain, France's new Prime Minister, sent a message to the Reich's leadership via Spanish ambassador to request a ceasefire, but the fighting went on. The remnants of the British force were evacuated from St. Nazaire using a flotilla of various vessels. In the afternoon the port was attacked by bombers of I. and II./KG 30, whose crews targeted the largest of the ships HMT *Lancastria*[35]. Officially, the ship carried 5,310 refugees, but they were mostly British troops. At around 16.00 the Ju 88s scored three direct hits on the ship dropping SC 500 bombs, which penetrated her hull below the waterline. The *Lancastria* first listed heavily and then capsized and sank after twenty minutes. 2,477 survivors were rescued, but 66 crew members, 2,833 British troops and several hundred civilian refugees, mainly French, perished. The sinking of the *Lancastria* was Britain's most tragic naval disaster of World War II.

During the same operation one of the German bombs hit SS *Oronsay* destroying the bridge, navigation room, wheel house and radio room. The ship's captain suffered a broken leg in the explosion. Several other vessels were also damaged by the Luftwaffe bombs. Luckily for the Allies, the onslaught was stopped by the French Morane MS 406 fighters, whose crews engaged and chased away the attacking Luftwaffe bombers. Uffz. Peter Wolfgang Stahl from II./KG 30 was involved in the fighting: "Over the target area I had my first encounter with the enemy fighters, the French Moranes. Initially they didn't attack me, as the fighting concentrated on the other end of the estuary. I saw one of the Ju 88s go down and egress the area at low level trailing black smoke from one of its engines. Moments later I saw the ships down below – a huge flotilla of freighters assembled at the mouth of the river. They were right there, no need to press on to St. Nazaire. As we approached the target we came under vicious AAA fire. I tried to stay slightly off to the side to have a better chance of picking a target. Suddenly Moritz reported: 'Fighters, closing in!'.

The crew of this Ju 88 A-1 from I./KG 51 are getting ready for another combat mission.

'How far are they?', somebody asked. 'There's only one, around 500 meters away!'.

I pushed hard on the yoke and went straight down without deploying the dive brakes. Moritz reported that the Morane was fallowing us down. We were diving at an incredible speed, 600 km/h, 680 just a moment later. Then I caught a large transport in my sight, pulled the nose up slightly and released the bombs. I then put the aircraft in a wide descending turn to take us away from the enemy flak. We were still going pretty fast, but Moritz and Jupp took turns to report the Morane was still behind us, unable to close in. I knew that the stubborn kid would get us as soon as I leveled off just above the ground. I had to think fast and come up with an evasive maneuver. A wide, steep turn to the right put some distance between us and the fighter. But the enemy followed suit. Using all the tricks in his bag and taking advantage of his superior maneuverability, he tried to assume the most favorable attack position, but I was still outside his range. And then the fighter...disappeared, almost as suddenly as it first popped up. It looked like he had enough of the workout my Ju 88 gave him and decided to head out to sea, where the battle was still raging, in search of an easier target. (...)

All in all, our squadron did exceptionally well. We scored lots of hits, all confirmed by other crews, with no losses at all. Uffz. Geffgen's aircraft was hit as many as 70 times, but

Junkers Ju 88 A photographed in France in the summer of 1940.

he still made it home safely. He made a textbook wheels-up landing back at the base using four 250 kg bombs as skids. He couldn't jettison the weapons, since both electric and mechanical jettison systems were out of action."[36]

A single Ju 88 A from I./KG 30 was also damaged during the raid. The machine, belonging to 3. Staffel, returned to base riddled by AAA shells and machine gun bullets, which killed Ofw. L. Edmüller. Fw. F. Erdel and Ogefr. H. Krauß suffered only minor wounds.

During the final day of fighting over France, on June 18, two Ju 88 A-1s from 9./LG 1 were shot down by AAA west of Brest. Among the crews were the CO of LG 1 Obst. Alfred Bülowius and commander of 9. Staffel Oblt. F.-W. Schaumann, Both German crews were captured by the French, but spent only three days in captivity. They were released after ceasefire had been announced on June 21, 1940.

Participation of the Ju 88 in the western campaign can hardly be considered a success, since some 80 aircraft (including reconnaissance machines) were lost in combat. The Ju 88's weakest link proved to be its insufficient defensive armament, which consisted of only three machine guns in flexible mounts. The problem was partly resolved by ad-hoc field modifications, which included mounting two additional guns in the canopy's rear side panels. There was also a provision for replacing a single gun in the rear canopy station (B-Stand) with two flexible guns, which increased the number of barrels to six.

Ju 88 A-1 of 1./KG 30 covered with camouflage netting. Note the squadron's insignia on the engine cowl.

Bibliography

Aufdemkamp Friedrich: Wir flogen mit dem Edelweiss, Waizendorf 2002
Baumbach Werner: „Zu Spät? Aufstieg und Untergang der deutschen Luftwaffe", Stuttgart 1978
Bekker Cajus: Angriffshöhe 4000, Ein Kriegstagebuch der deutschen Luftwaffe, Oldenburg 1964
Brütting Georg: Das waren die deutschen Kampflieger-Asse 1939-1945, Stuttgart 1977
Cornwell Peter D.: The Battle of France, Then and Now, Six Nations Locked in Aerial Combat September 1939 to June 1940, Old Harlow 2007
Cull Brian, Lander Bruce, Weiss Heinrich: Twelve Days in May, London 1995
Dierich Wolfgang: Kampfgeschwader 51 „Edelweiss", Eine Chronik aus Dokumenten unf Berichten 1937-1945, Stuttgart 1991
Gundelach Karl: Kampfgeschwader „General Wever" 4, Stuttgart 1978
Herrmann Hajo: Bewegtes Leben, Kampf- und Jagdflieger 1935-1945, München 1993
Janowicz Krzysztof: Junkers Ju 88, vol. II, Lublin 2005
Kohl Hermann: Wir fliegen gegen England, Einsatz der Luftwaffe 1939/1940, Reutlingen 1941
Medcalf William A.: Junkers Ju 88, Volume One: From Schnellbomber to Multi-Mission Warplane, Development, Production, Technical History, Hersham 2013,
Murawski Marek J.: Luftwaffe – Działania bojowe, tom 1, Warszawa 1998
Nowarra Heinz J.: Die Ju 88 und ihre Folgemuster, Stuttgart 1981
Piekalkiewicz Janusz: Luftkrieg 1939-1945, München 1982
Shores C., Ehrengardt C.-J., Foreman J., Weiss H., Olsen B.: Fledgling Eagles, London 1991
Šnajdr Miroslav: Pád Paříže, Letecká válka v deseti červnových dnech roku 1940, Olomouc 2000,
Spick Mike: Asy lotnictwa bombowego Luftwaffe, Warszawa 2006
Stahl Peter W.: Kampfflieger zwischen Eismeer und Sahara, Stuttgart 1990
Weal John: Ju 88 Kampfgeschwader on the Western Front, Oxford 2000

Endnotes

[1] Ju 88 experimental unit (author's note).
[2] Dipl.-Ing. Carl Francke was an aviator and test pilot. In 1937 Francke and Ernst Udet flew Bf 109s during Zurich Flying Meet. Before the war he worked at Erprobungsstelle Rechlin, where he took part in the Ju 88 flight test program. Later on Francke served with Erprobungskommando 88, 1./KG 25 and I./KG 30. In November 1939 he became a Heinkel factory test pilot and flew the He 177 on its maiden sortie. In 1941 Francke became head of E-Stelle Rechlin. He survived the war (author's note).
[3] Kohl Hermann: Wir fliegen gegen England, Einsatz der Luftwaffe 1939/1940, Reutlingen 1941, pp.. 40-41.
[4] Weal John: Ju 88 Kampfgeschwader on the Western Front, Oxford 2000, p. 12.
[5] Bekker Cajus: Angriffshöhe 4000, Ein Kriegstagebuch der deutschen Luftwaffe, Oldenburg 1964, p. 73-75.
[6] Ibidem, op. cit. pp. 77-79.
[7] Spick Mike: Asy lotnictwa bombowego Luftwaffe, Warszawa 2006, p. 42.
[8] Ibidem, op. cit. p. 43.
[9] Zirkus Renz – one of the oldest German circuses, established in 1842 in Berlin (author's note).
[10] Kohl Hermann..., op. cit. pp. 62-63.
[11] Ibidem, op. cit. pp. 64-66.
[12] HMS *Sphinx* – displacement: 1,372 tons, length: 74.9 m, beam: 10.21 m, draft: 2.7 m, speed: 16.5 knots, range: 7,200 nm at 10 knots, armament: two 102 mm guns, 4 12.7 mm machine guns, 4 7.7 mm machine guns, 4 depth charge throwers, 2 depth charge rails (author's note).
[13] Kohl Hermann..., op. cit. p. 111.
[14] Ibidem, op. cit. 112-113.
[15] HMS *Norfolk* was a *County* class battlecruiser launched on December 12, 1928 and commissioned on April 30, 1930. Full displacement – 13,640 tons; length – 192.86 m; beam – 20 m; draft – 5.5 m; maximum speed – 31.5 knots; range at 12 knots – 12,000 nm; complement – 819; armament – 8 203 mm guns (4xII), 8 102 mm guns (4xII), 16 40 mm cannons (2xVIII), two Supermarine Walrus flying boats (author's note).
[16] HMS *Gurkha* was a *Tribal* class destroyer launched on July 7, 1937 and commissioned on October 21, 1938. The ship displaced 2,519 tons, measured 114.9 m in length, had a beam of 11.2 m. and draft of 3.96 m. Her maximum speed was 36 knots and at 15 knots she had a range of 5,700 nm. Armament: 8 120 mm guns (4xII),4 40 mm cannons (1xIV), 8 12.7 mm machine guns (2xIV), 4 533 mm torpedo tubes (1xIV) and 2 depth charge throwers. Complement: 190 officers and enlisted personnel (author's note).
[17] HMS *Suffolk* was a *Kent* class battlecruiser launched on February 16, 1926 and commissioned on May 31, 1928. Length overall: 190.0 m; beam: 28.8 m; draft: 4.95 m; displacement: 9,750 tons; maximum speed: 31.5 knots; armament: 8 203 mm guns (4xII), 8 102 mm guns (8xI), 4 40 mm cannons (4xI), 8 12.7 mm machine guns (2x4), 8 533 mm torpedo tubes (2xIV). Complement: 700 officers and enlisted men (author's note).
[18] Baumbach Werner: „Zu Spät? Aufstieg und Untergang der deutschen Luftwaffe", Stuttgart 1978, pp. 101-103.
[19] *Trygg* was a Royal Norwegian Navy torpedo boat displacing 256 tons. The ship measured 53 m in length, had a beam of 5.5 m and draft of 1.58 m. She was capable of a top speed of 25 knots and carried a crew of 33. Armament: 1 76 mm gun, 2 20 mm cannons, 2 machine guns and 2 torpedo tubes (author's note).
[20] Stahl Peter Wolfgang: Kampfflieger zwischen Eismeer und Sahara, In meinem Fall: Ju 88, Stuttgart 1985, pp. 30-31.
[21] Baumbach..., op. cit. pp. 106-108.
[22] Cull Brian, Lander Bruce, Weiss Heinrich: Twelve Days in May, London 1995, p. 26.
[23] Aufdemkamp Friedrich: Wir flogen mit dem Edelweiss, Waizendorf 2002, pp. 102-104.
[24] Stahl..., op. cit. p. 34.
[25] Aufdemkamp..., op. cit. pp. 104-106.
[26] Ibidem, op. cit. p. 106.
[27] Kaczmarek – a Luftwaffe jargon for a wingman in a flight or a two-ship formation (author's note).
[28] Dierich Wolfgang: Kampfgeschwader 51 „Edelweiss", Eine Chronik aus Dokumenten unf Berichten 1937-1945, Stuttgart 1991, pp. 85-89.
[29] Herrmann Hajo: Bewegtes Leben, Kampf- und Jagdflieger 1935-1945, München 1993, pp. 88-89. The ship was probably the French freighter *Orphelia* (6,477 tons).
[30] *Sirocco* – displacement: 1,68 tons; length: 106.00 m; beam: 4.3 m; maximum speed: 33 knots; range: 2,150 nm at 14 knots; complement: 7 officers and 138 enlisted men; armament: 4 130 mm guns (4xI), 2 37 mm cannons (4xI), 6 550 mm torpedo tubes (2xIII).
[31] Tommy – Luftwaffe jargon for British troops (author's note).
[32] Herrmann..., op. cit. pp. 89-91.
[33] Stahl..., op. cit. pp. 41-42.
[34] Ibidem..., op. cit. pp. 43-44.
[35] RMS *Lancastria* was a British liner operated by Cunard White Star. The ship was launched on May 31, 1920 and began service on June 13, 1923. The 16, 243 BRT vessel measured 176.00 m in length and 21.00 m in beam. She had a draft of 11.9 m and a top speed of 16.5 knots. *Lancastria* could accommodate 1,785 passengers. In April 1940 she was commissioned as a troop carrier HMT (His Majesty's Transport) *Lancastria* (author's note).
[36] Stahl..., op. cit. pp. 46-47.

A small, standby compass used by the bombardier was mounted on the framing of the bomber's glazed nose. The compass was mounted in a small, semi-openwork box.

The compass mounted on the framing of the nose glazing.

The cockpit's rear wall featuring radio equipment panels and spare MG 15 magazines.

Two views of stowed ammunition magazines.

MG 15 magazines stowed on the lower part of the cockpit's rear wall.

Spare magazines stowage brackets.

Spare MG 15 magazines stowage brackets on the cockpit's rear wall.

BZG-2 (Bombenzielgerät 2) bombsight was mounted in the forward fuselage, just above the ventral gondola.

Another view of the BZG-2 bombsight.

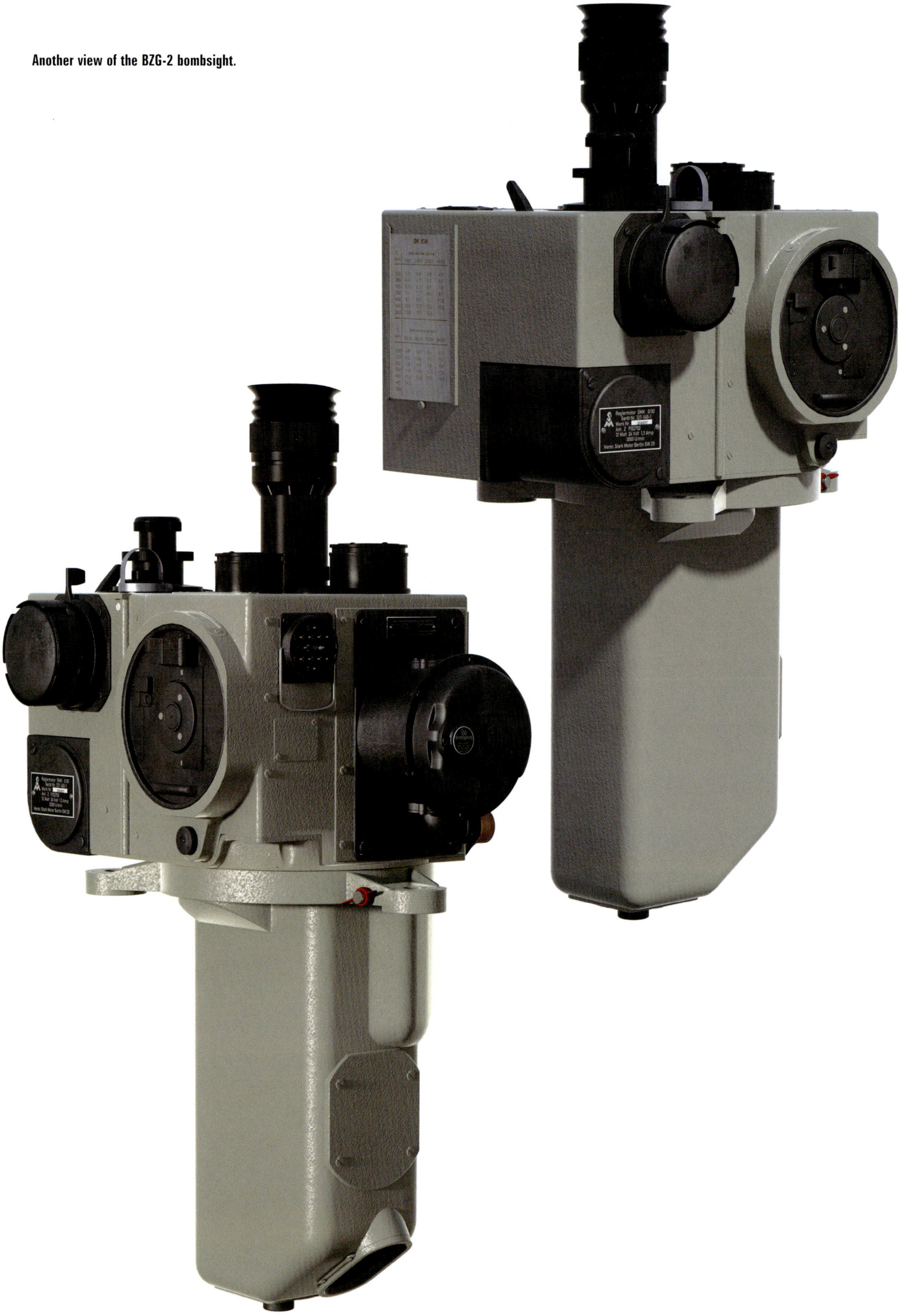

Bottom part of the BZG-2. The shape of the metal casing covering the optics is clearly visible.

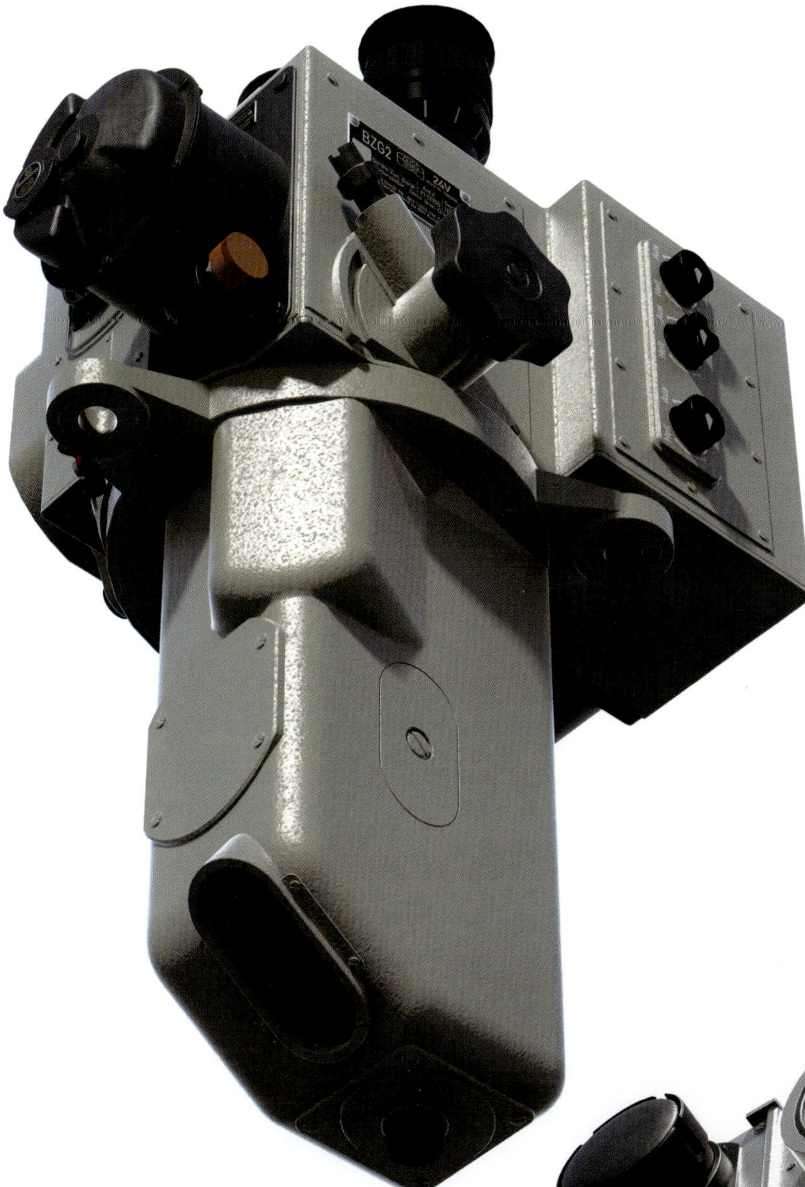

Bottom part of the BZG-2 column with a glazed viewfinder.

Cover of the electric motor of the BZG-2 bombsight.

Details of the front, and right-side panels of the BZG-2 bombsight.

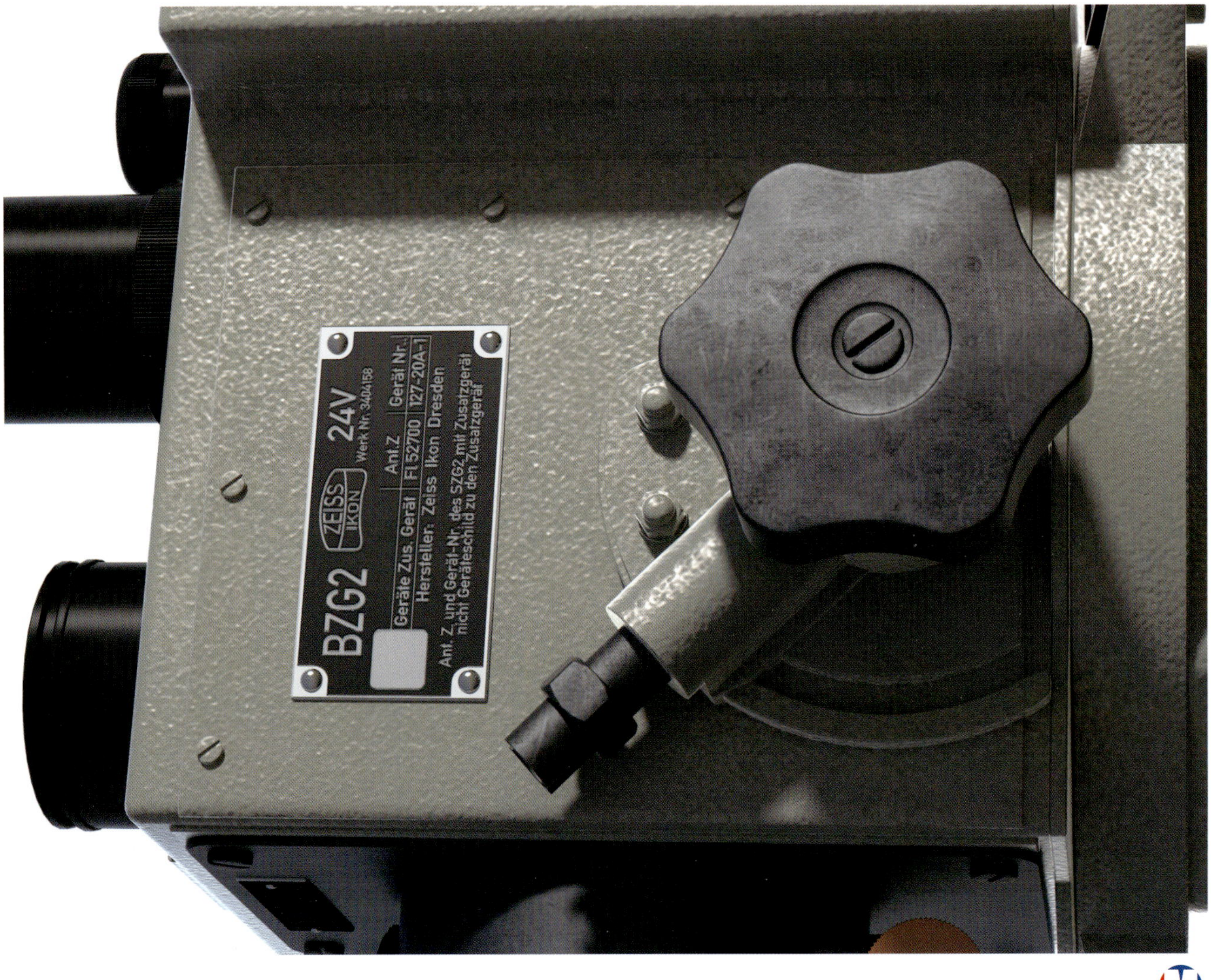

Details of the BZG-2.

Top plate of the BZG-2.

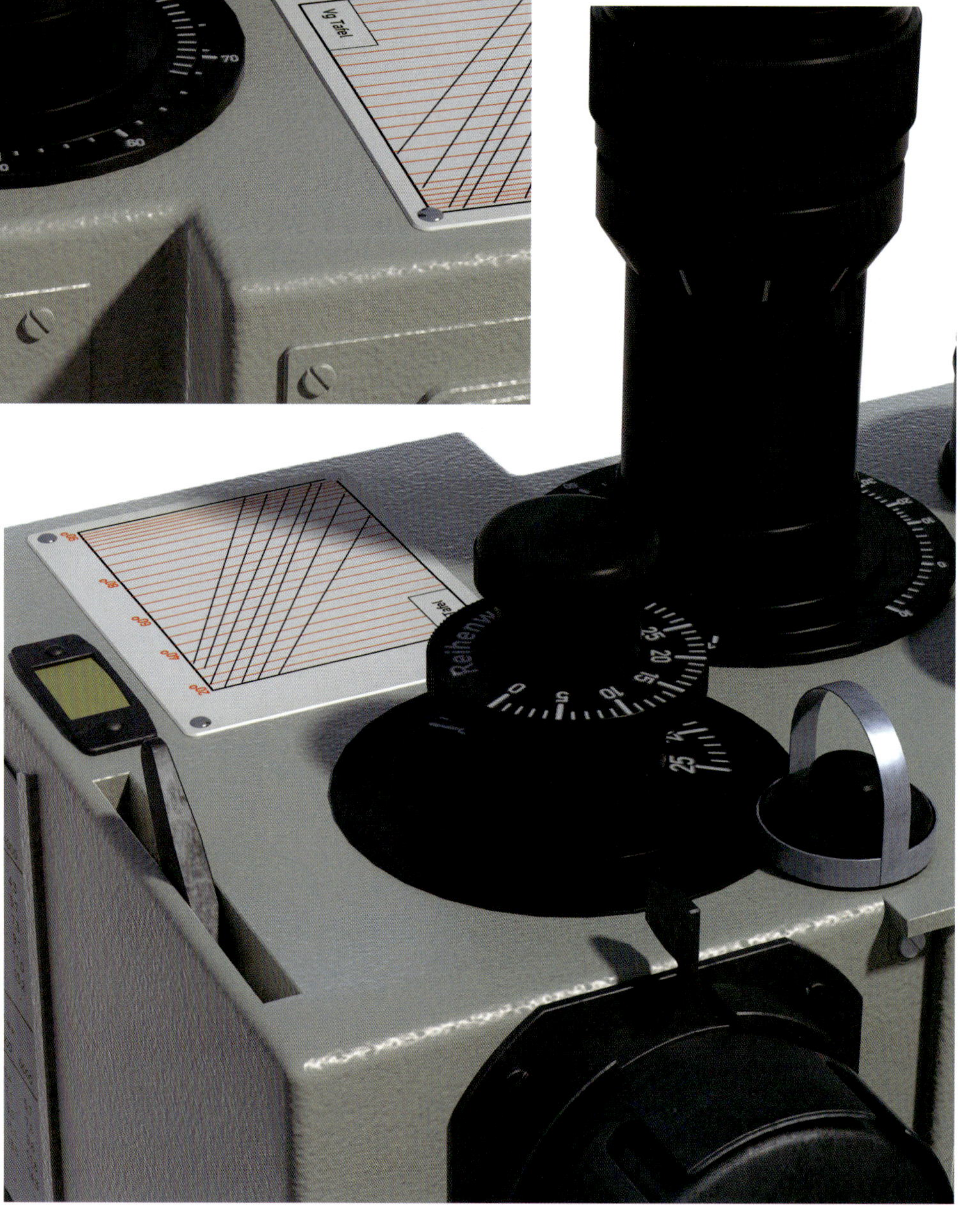

Details of the top plate of the BZG-2.

Details of the top plate of the BZG-2.

Location of the BZG-2 relative to the ventral gondola.

The cockpit floor with the BZG-2 installation above the ventral gondola.

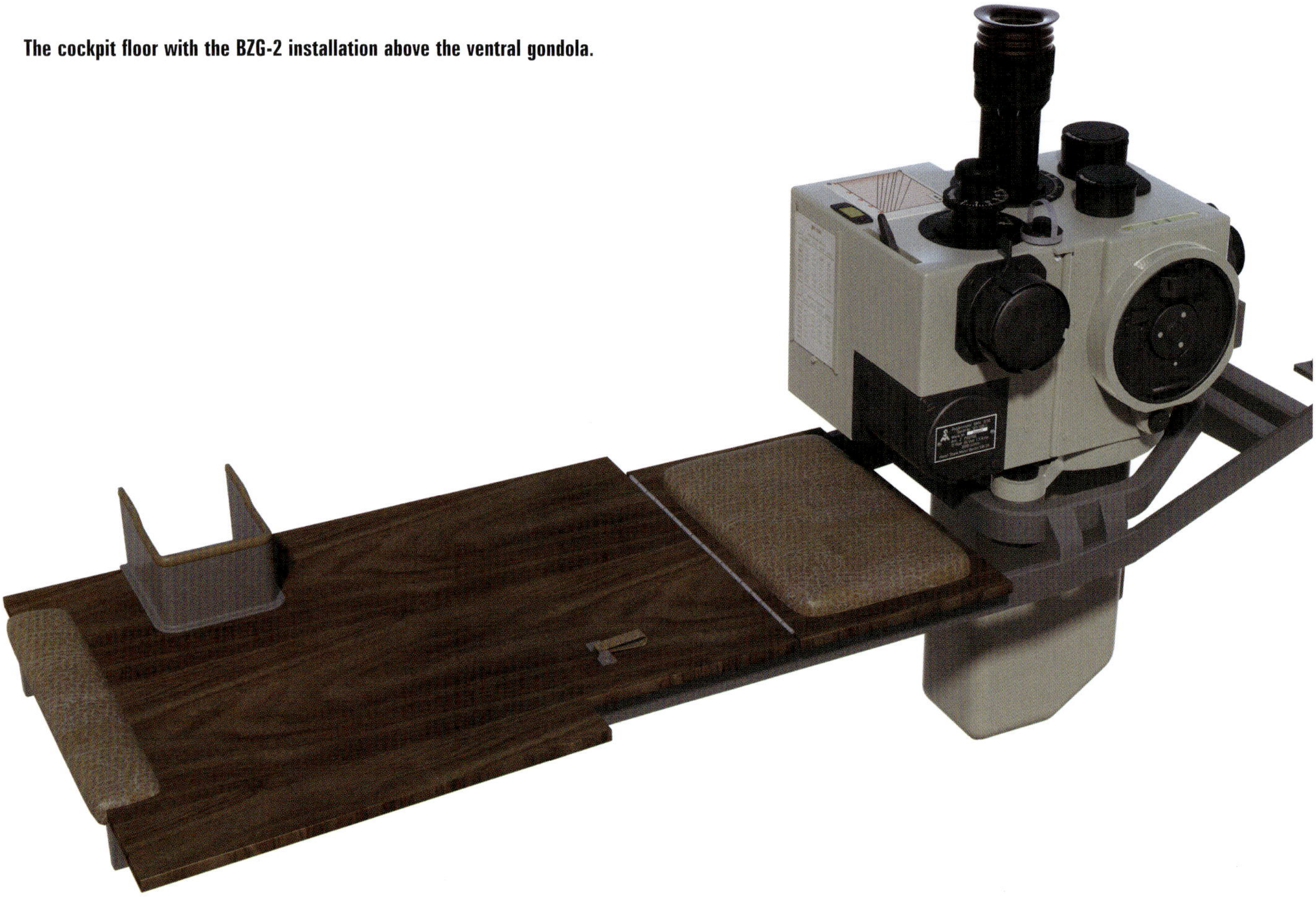

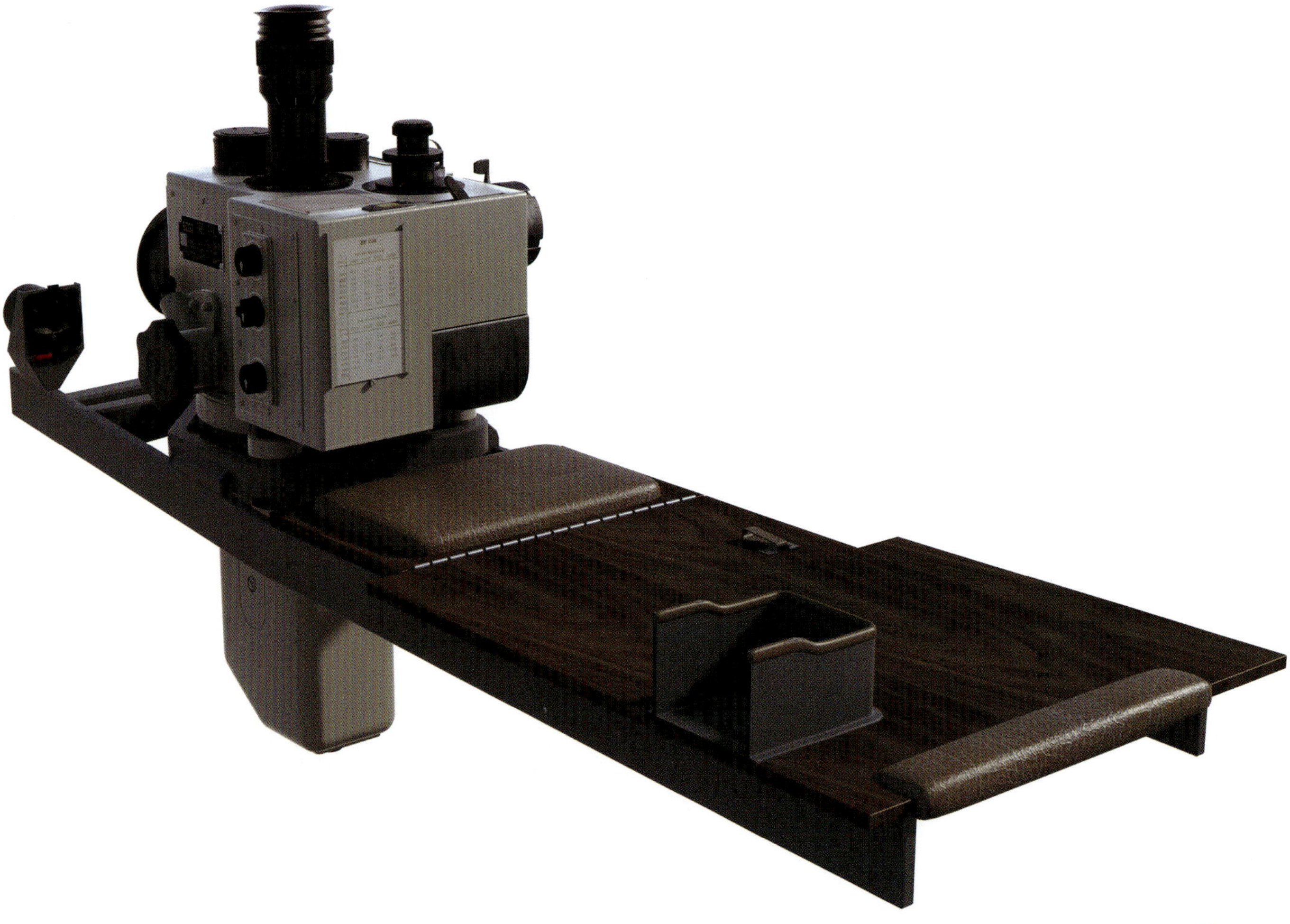

Details of the BZG-2. Clearly visible is the turn indicator.

Details of the BZG-2 mounted on the cockpit floor above the ventral gondola.

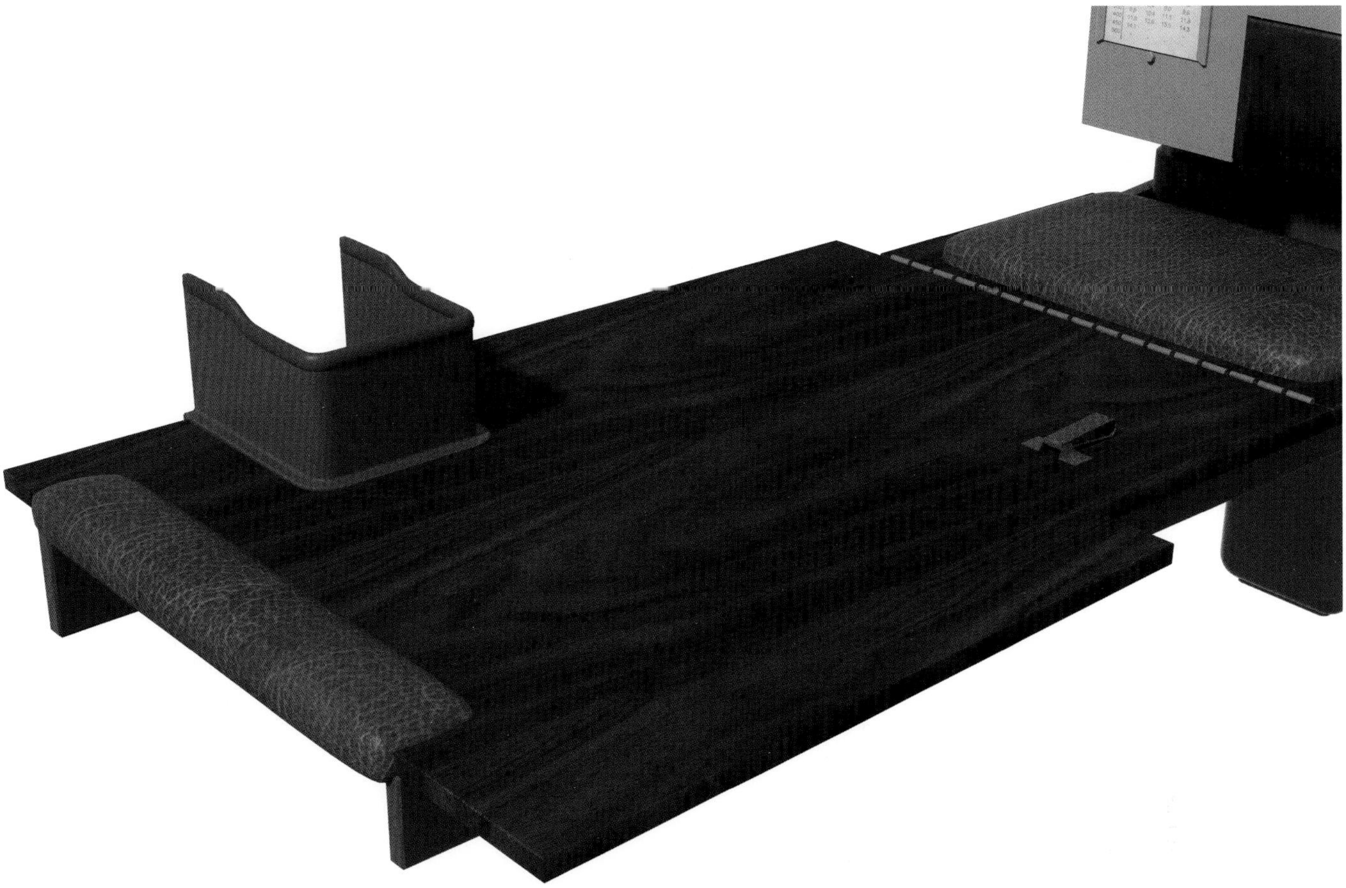

Details of the BZG-2 mounted on the cockpit floor above the ventral gondola.

The complete instrument panel, including Revi 12C/d sight in a mounting bracket and the main compass.

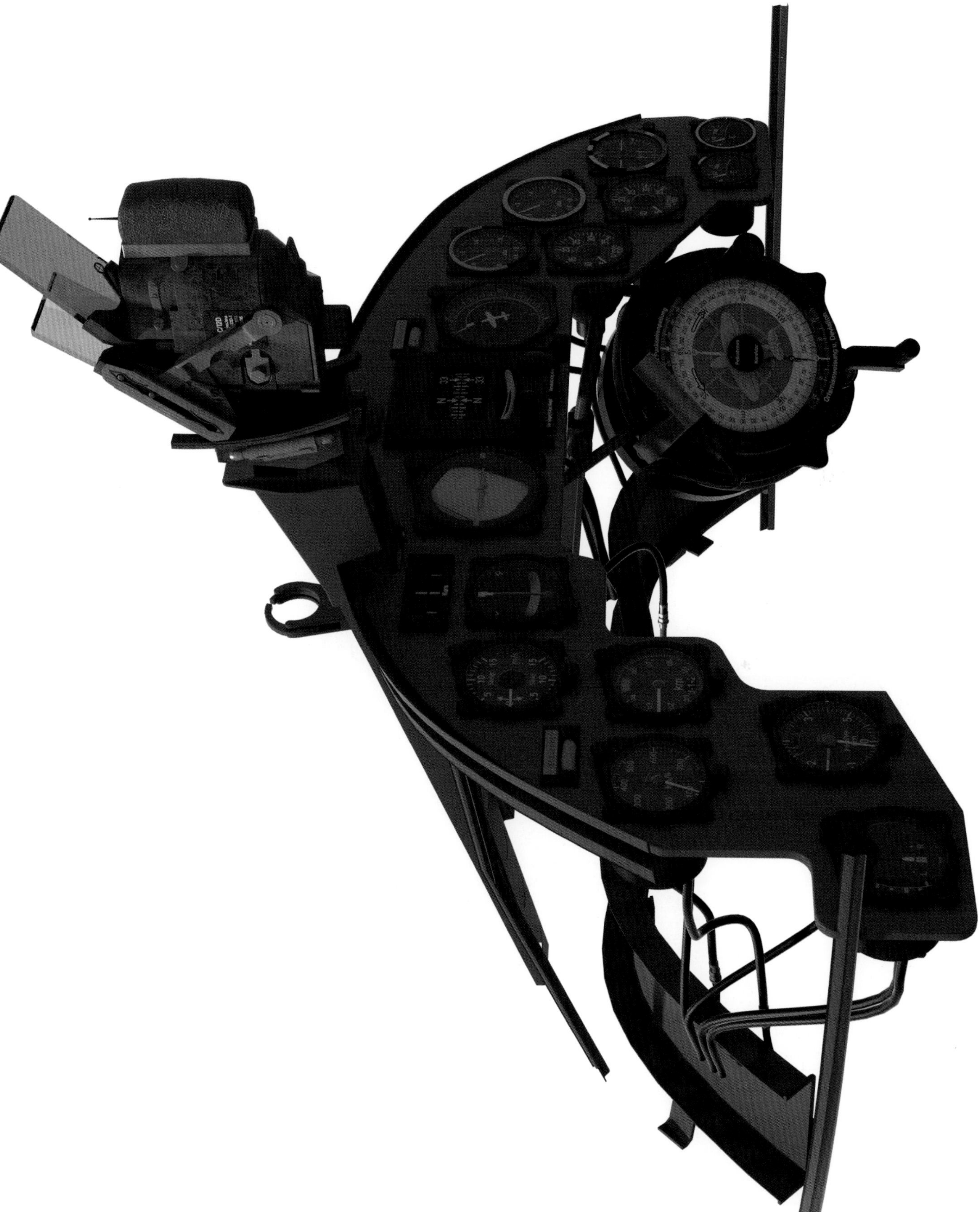

The complete instrument panel, including Revi 12C/d sight in a mounting bracket and the main compass.

Compass and a flash light on the instrument panel. The metal bar with a wooden spherical knob was used to unlock the forward-firing MG 15 machine gun.

Details of the Revi 12C/d in its mounting bracket.

The locking bracket on the barrel
of the forward MG 15 machine gun.
The bracket's location relative to the
instrument panel is clearly visible.

A view from below showing the rear of the instrument panel. The wiring has been simplified for clarity.

The locking bracket of the forward MG 15 machine gun. In the unlocked position the bracket was stowed under the forward fuselage skin panels in front of the windshield.

The main directional gyro in the Ju 88 was mounted under the instrument panel, on the right-hand side.

The main directional gyro in the Ju 88 was mounted under the instrument panel, on the right-hand side.

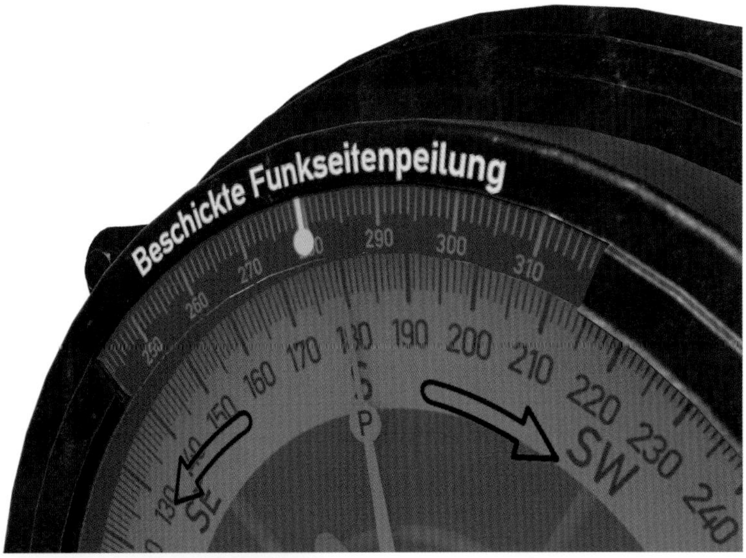

Details of the main gyro.

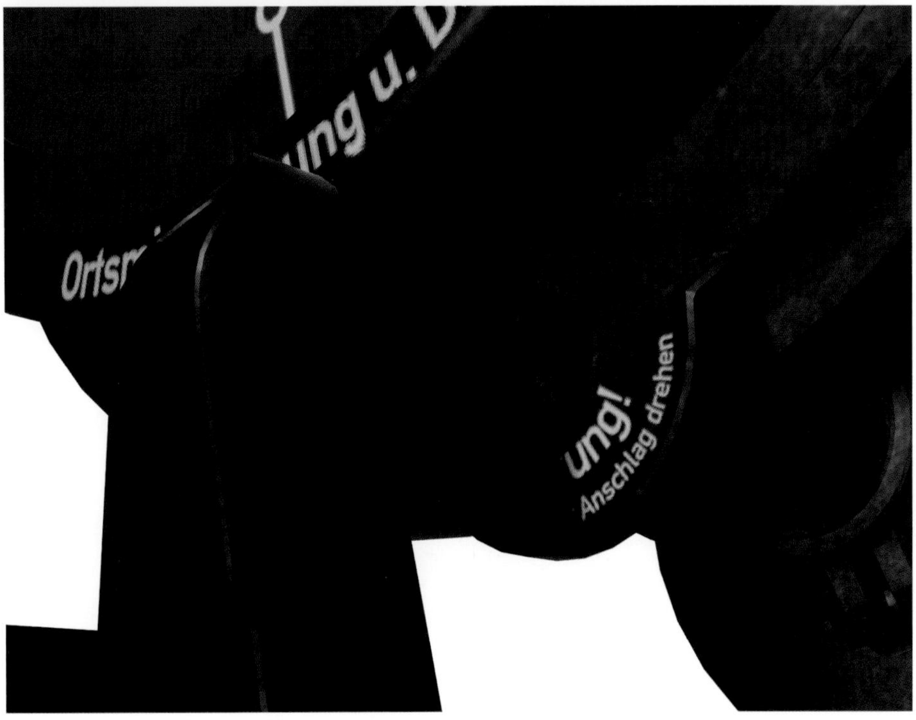

Main instrument panel in a production Ju 88A-1.

Details of the main instrument panel: on the left is a vacuum-driven Askania (Fl.22426) artificial horizon, while Siemens (Fl.22561) heading indicator is in the middle.

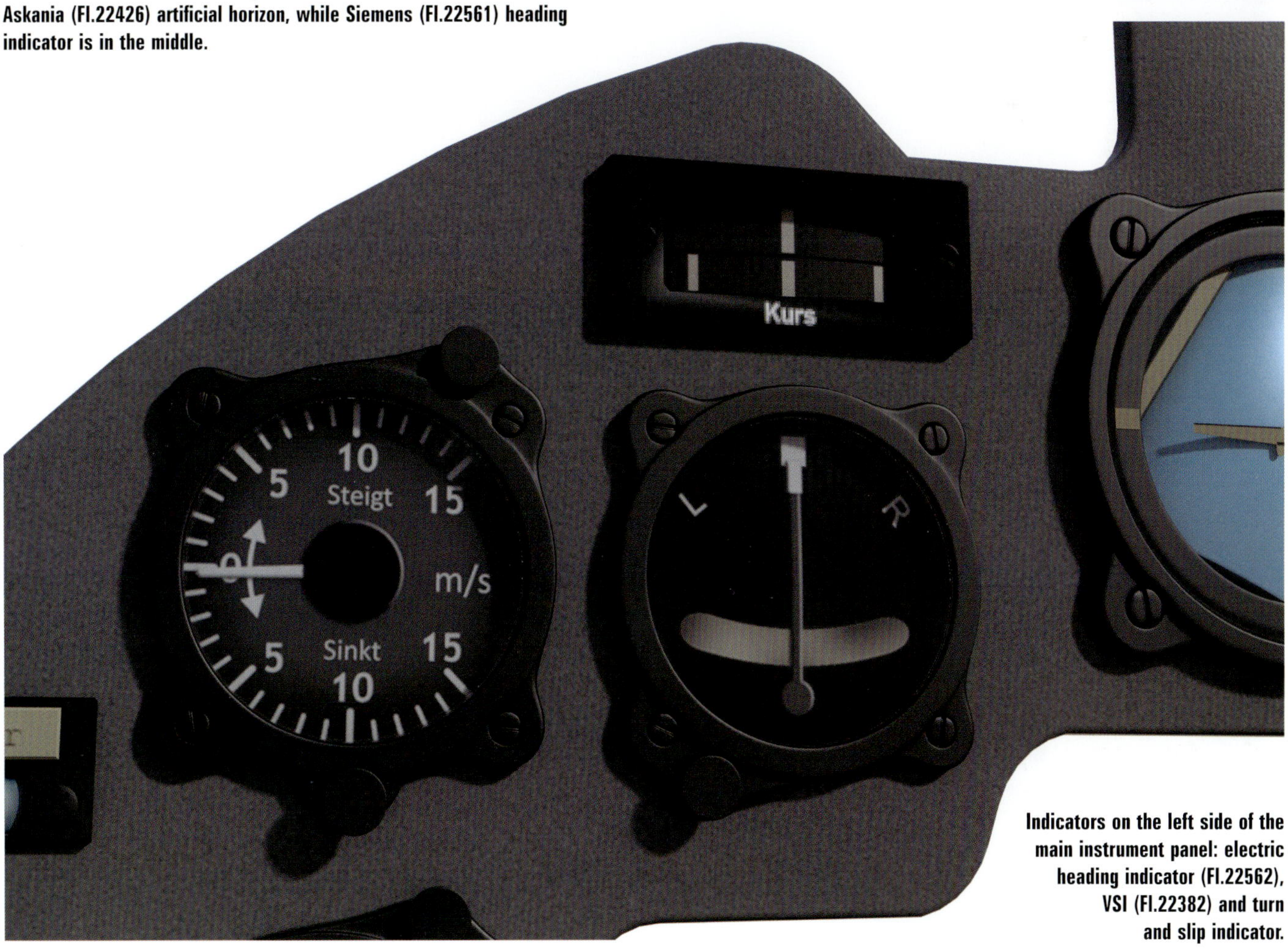

Indicators on the left side of the main instrument panel: electric heading indicator (Fl.22562), VSI (Fl.22382) and turn and slip indicator.

Sequenced bomb release unit RAB 14d.

Details of the RAB 14d unit.

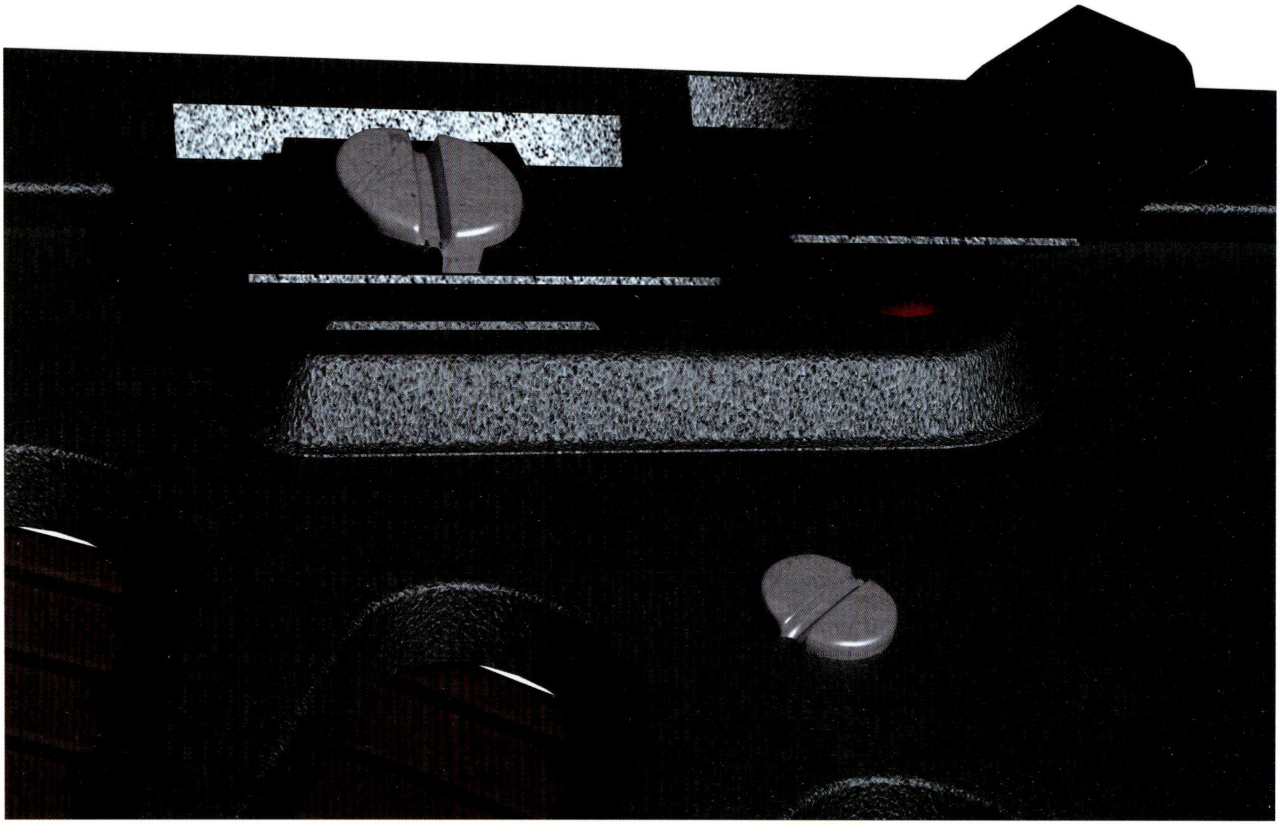

Details of the RAB 14d unit.

Details of the RAB 14d unit.

Rudder pedals.

The rudder bar's design allowed adjustment of rudder pedals to fit pilots of different heights. The illustration shows to good advantage the hydraulic brakes system (the pedals featured main wheel braking).

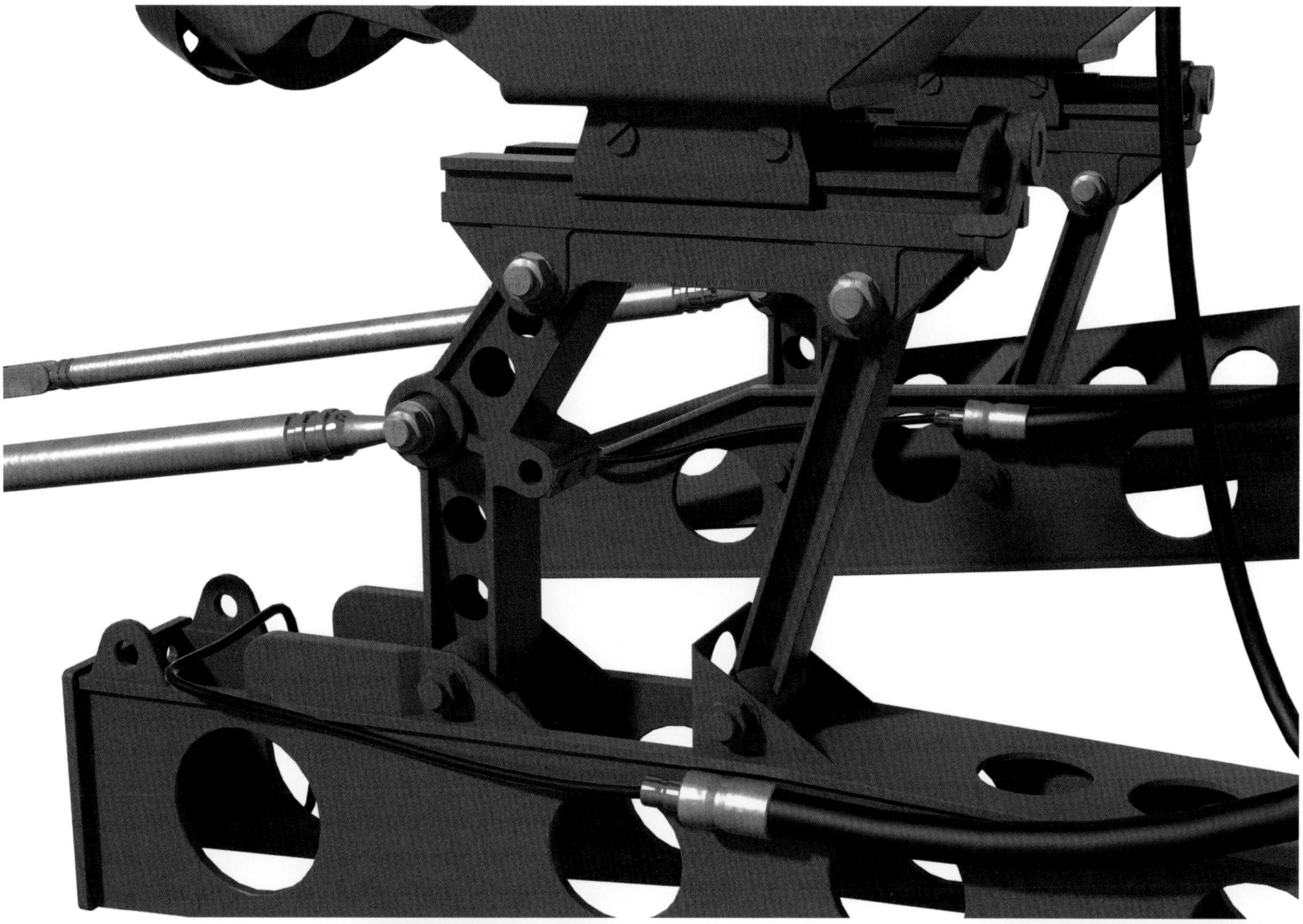

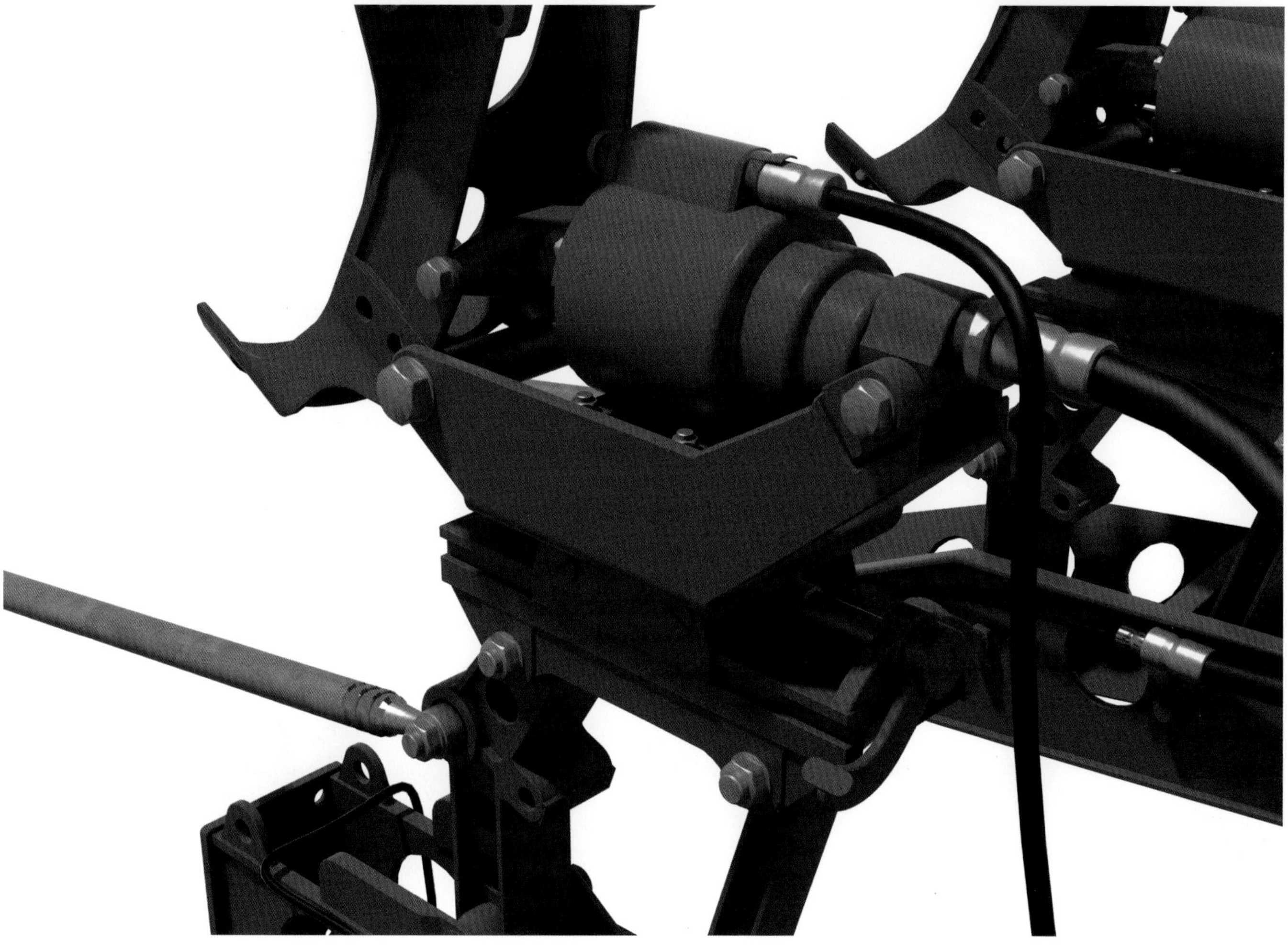

A full set of flight controls, including rudder bar and pedals and control column.

Details of the control column mounting and foot-operated adjustment mechanism of rudder pedals.

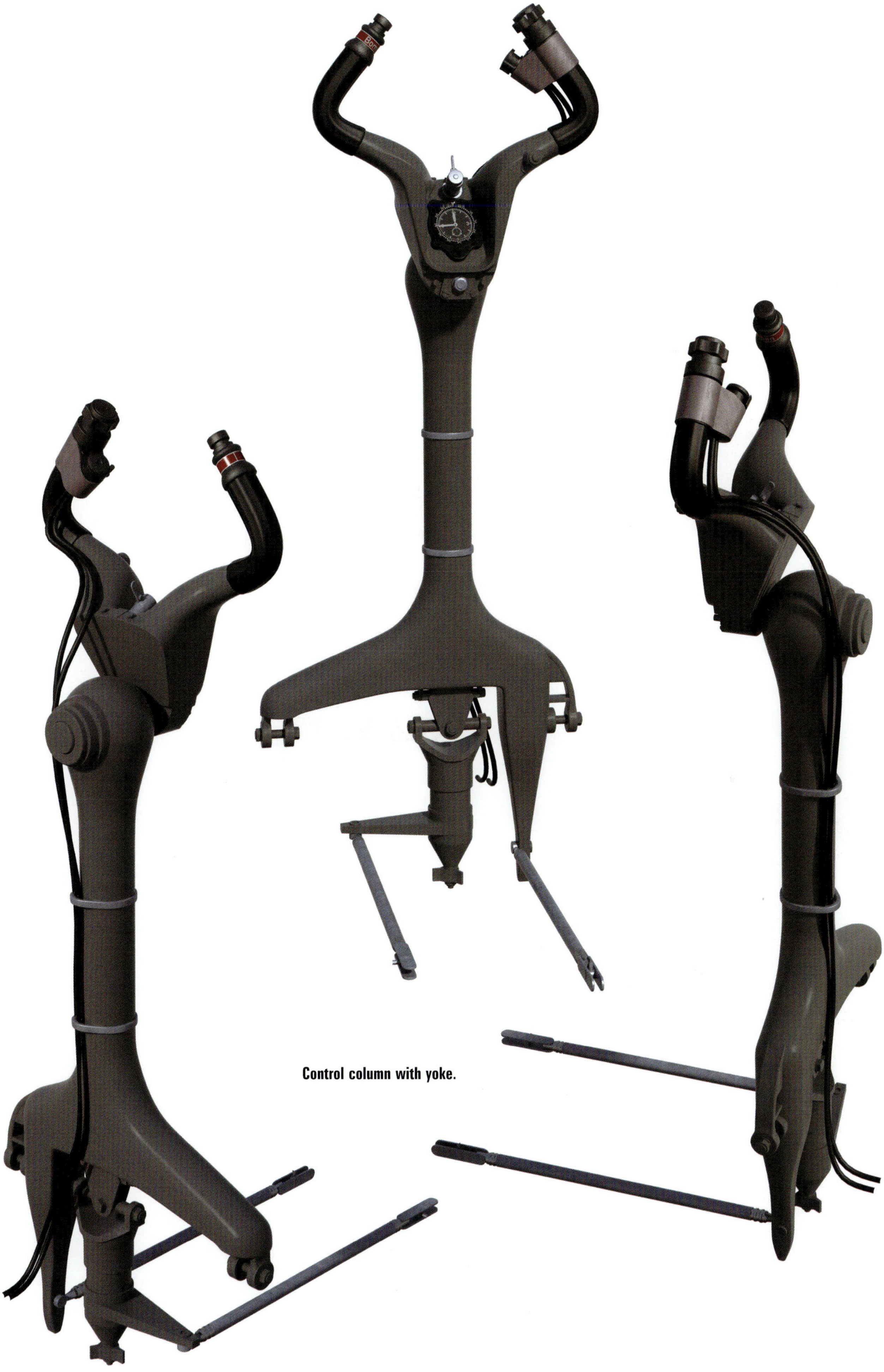

Control column with yoke.

A Ju 88 yoke with a clock mounted in the middle.

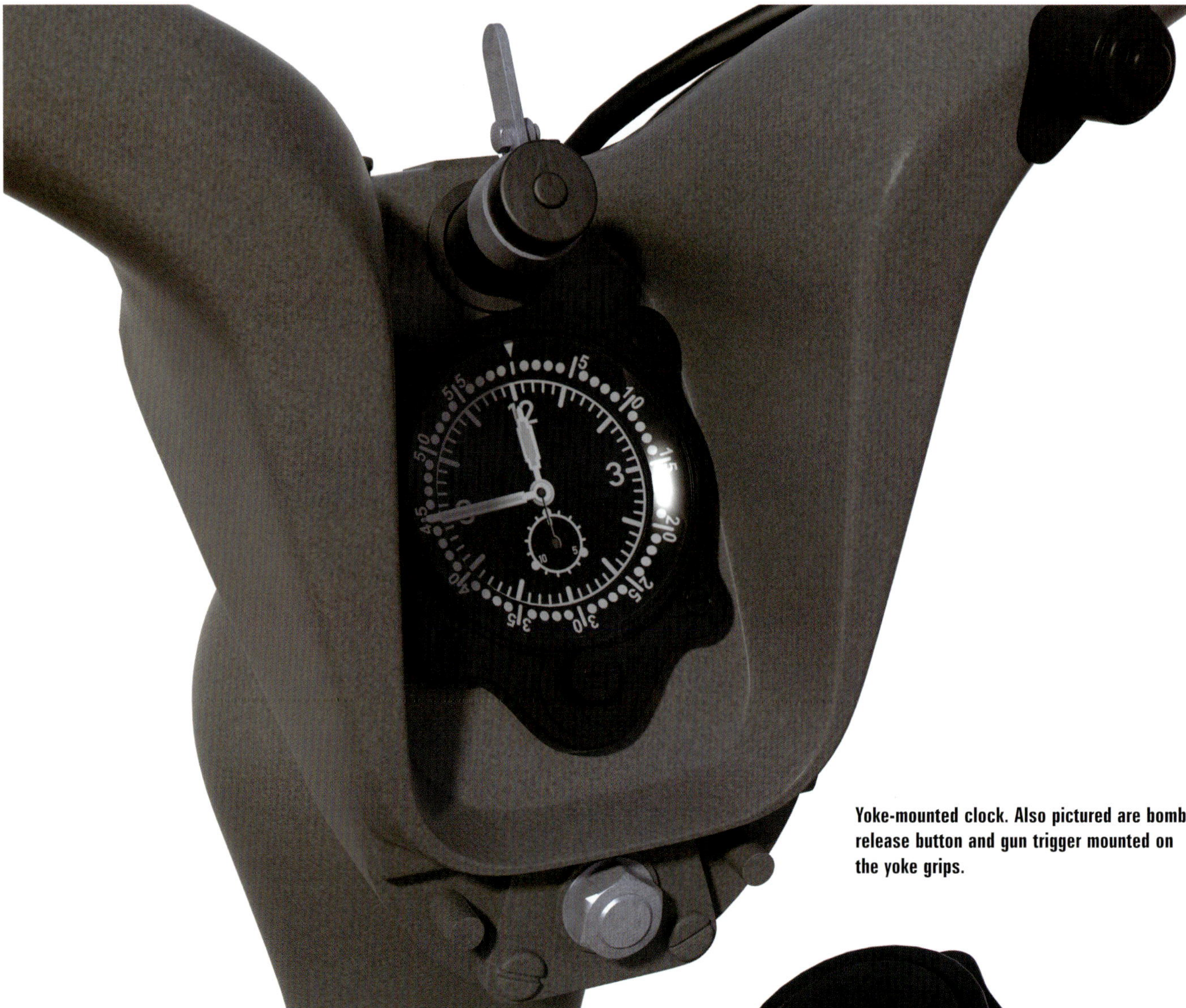

Yoke-mounted clock. Also pictured are bomb release button and gun trigger mounted on the yoke grips.

A Ju 88A-1 undergoing maintenance – bottom part of the fuselage and the ventral gondola.

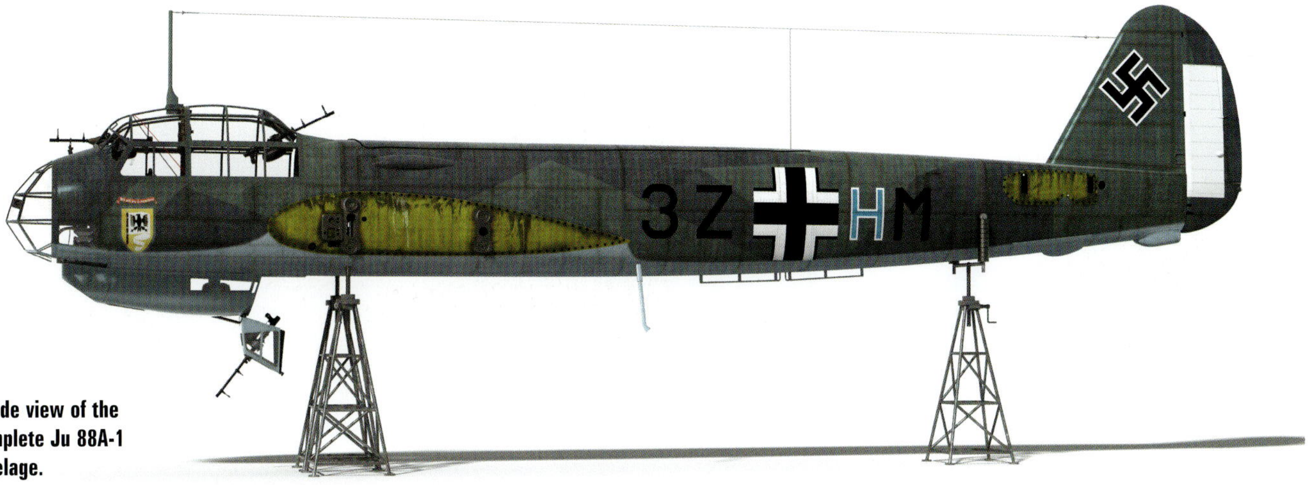

A side view of the complete Ju 88A-1 fuselage.

A characteristic blister on the fuselage skin just aft of the cockpit.

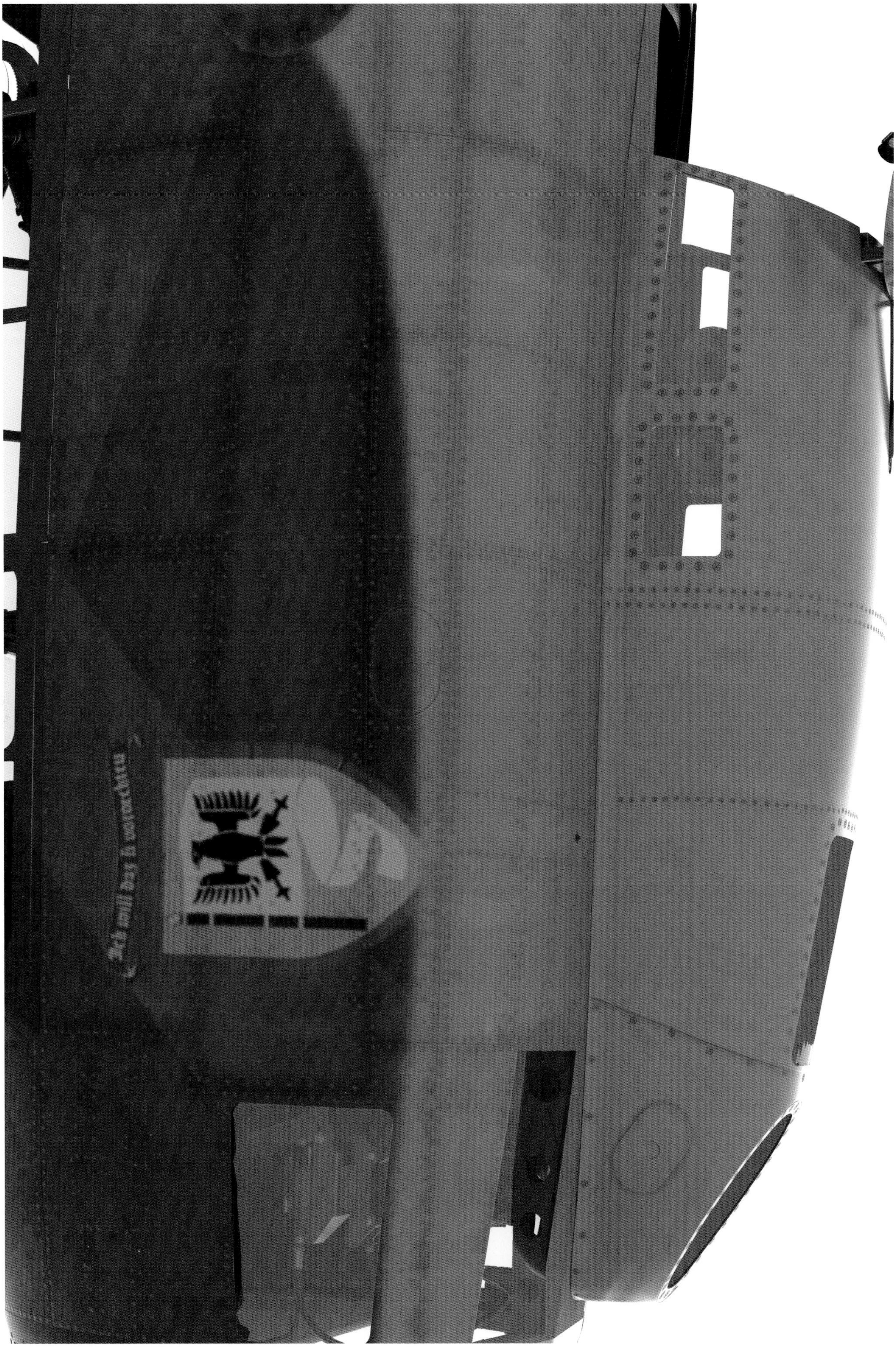

Skin panels covering the forward fuselage section.

A fragment of fuselage skin behind the cockpit. Clearly visible are fuselage fuel tanks filler covers.

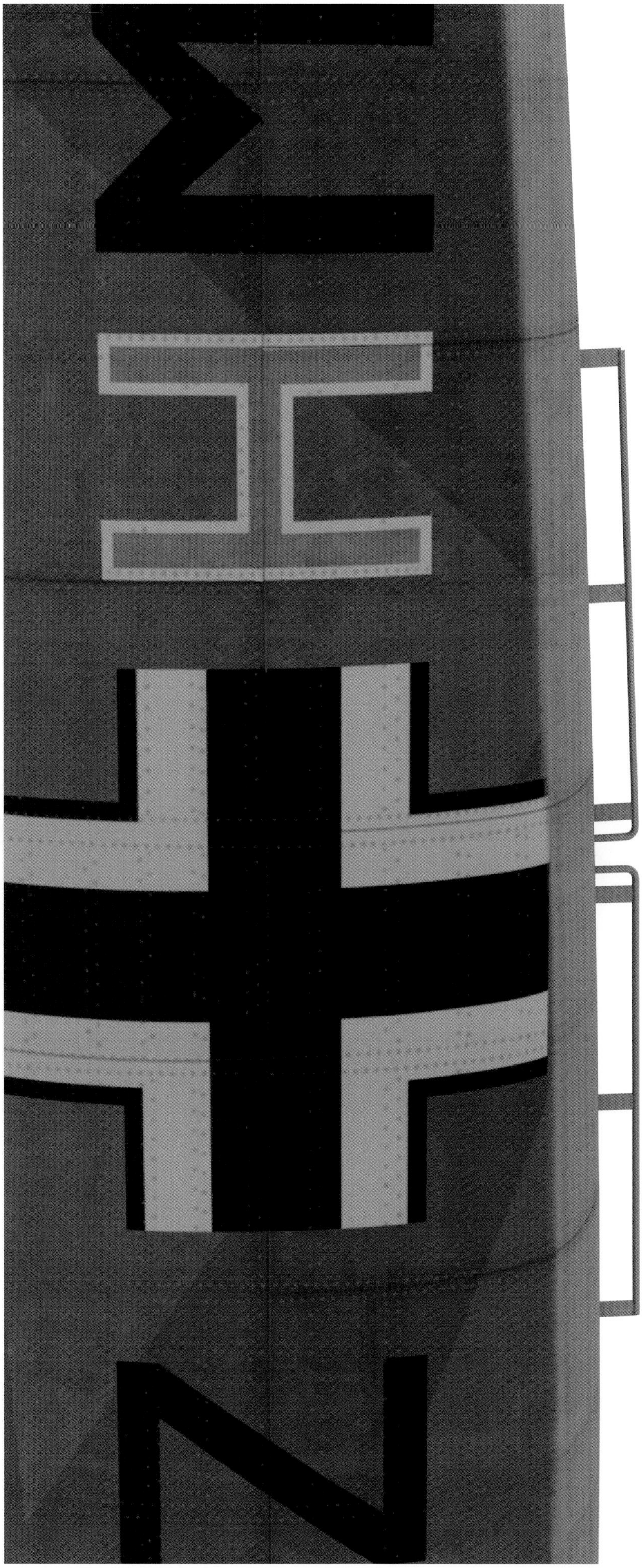

Antennas of the FuBL 2 "blind landing" system.

A view from the rear showing skin panels covering the fuselage dorsal section.

Characteristic sheet metal fairings streamlining the canopy-fuselage connection, in this case the canopy with a single machine gun mount in the rear.

A view of the ventral tub hatch in the open position.